OLDENBOURG'S PRACTICAL CHARTS

TABLES FOR HEAT ENGINEERS

THE HEATING OF ROOMS

COMPRISING FORTY CHARTS AND TABLES
FOR SIMPLYING CALCULATIONS

WITH EXPLANATION IN THREE LANGUAGES
ENGLISH - GERMAN - FRENCH

LONDON

THE TECHNICAL PRESS LTD

5 AVE MARIA LANE, LUDGATE HILL, E.C.4.

1936

VERLAG VON R. OLDENBOURG, MÜNCHEN UND BERLIN

Preface

This collection of tables is intended to assist the heating engineer in his calculations and in the preparation of projects. With the assistance of these tables, a large number of constantly recurring calculations can be carried out very quickly and yet with a sufficient accuracy; the reading of the values from the tables will be the quickest method in every case. But apart from that, the tables have a great value as they show everything at a glance, and give graphic illustration of the numerical values occuring in the calculations. It is not therefore surprising that a number of diagrams (such as for instance the well known IS-diagram) have been already adopted in similar form in various branches of industry; such diagrams enable the engineer to carry out and to follow his calculations one may say in the same way as a general follows his battles on a map.

Calculation tables are not however intended to be, and cannot be, a teaching manual; it must be assumed that their user is thoroughly familiar with the values and with the method of calculating. For each problem of calculation, the basic or starting values must be known, either from measurements or from estimation. But knowing them, the heating engineer will find ample assistance in the calculation tables for simplifying and reducing his usual work of calculation. In addition, they make practicable many calculations where formerly he could only estimate in a very approximate manner. The tables thus become an effective time and labour-saving device for the heating engineer who, compared with other engineers, hitherto greatly lacked such devices. But like in the case of any other tools, it is necessary to learn its use and to acquire practice so that the heating engineer should gradually come to utilise one table after another in the work daily required of him. To enable him to do so, short instructions for the use of the tables are given here.

The value of the calculation tables will be fully appreciated only when after a certain period of time (say a month), the engineer will realise the amount of time which he succeeded in saving by the use of this book. It must be however admitted at once that the tables could be improved, and the publishers will welcome therefore any suggestions.

HOW TO USE THE TABLES

1. Before using the tables for practical work, one should become familiar with the use of every individual table, with the calculation values which are known and sought, and with the following of the values in the table.

2. The simplest way to do so is by following the numerical example given on the text page, to which correspond the chain lines with the arrows in the calculation table. The agreement of the figures given in the text with the corresponding division strokes shows at once how the table is to be used also for any other values.

3. Begin for instance with the simplest table 16 which gives the most important nominal pipe dimensions. The example gives for the nominal diameter $d_n =$ 100 mm, the corresponding values of the inside diameter d_i, the thickness of wall δ, inside cross-section F_i, volume V_i, iron cross-section F_E and iron weight G_E. The value given in the text will be found at once by following the chain line and passing from the line marked with the reference letter (for instance d_i) to the marginal division with the same reference letter.

4. To increase the range of the tables, in the cases which can be recognised from the attached formula as being proportional, the corresponding values could be multiplied by any desired power of 10 as indicated in the individual examples (Tables 5, 7, 11, 18, 21, 23, 36).

Inhaltsverzeichnis.
Table of Contents.
Table des Matières.

1

5

6

IV. Bemessung der Rohrnetze.
Design of Piping.
Calcul de la tuyauterie.

V. Heizvorrichtungen.
Heating Equipment.
Radiateurs et échangeurs de chaleur.

Abgekürzte Quellenangabe.

Rietschel = H. Rietschels Leitfaden der Heiz- und Lüftungstechnik, 10. Aufl.
Berlin 1934.
Recknagel = H. Recknagels Hilfstafeln zur Berechnung von Warmwasser-
heizungen. 6. Aufl. München 1933.
DIN 4701 = Regeln für die Berechnung des Wärmebedarfs und der Heizkörper-
und Kesselgrößen von Warmwasser- und Niederdruckdampf-Heizungs-
anlagen (Dinorm 4701). Berlin 1929.
Hütte = Hütte, 26. Aufl., Bd. I, Berlin 1932.

Zeichenerklärung.

Key to Symbols.

Explication des notations.

9

i_D	$\dfrac{kcal}{kg}$	Wärmeinhalt des Dampfes . . . . Heat content of steam Chaleur de la vapeur	40
i_W	$\dfrac{kcal}{kg}$	Wärmeinhalt des Wassers . . . . 18, 19, 40 Heat content of water Chaleur contenue dans l'eau	
k	$\dfrac{kcal}{m^2\,h\,^oC}$	Wärmedurchgangszahl . . . . 5, 6, 31, 32 Coefficient of heat transmission Coefficient de transmission	
k'	$\dfrac{kcal}{m^2\,h\,^oC}$	Teil-Wärmedurchgangszahl . . . . Partial coefficient of heat transmission Coefficient de transmission partiel	5
k_B	$\dfrac{M}{t}$	Brennstoffkosten . . . . Fuel costs Prix du combustible (par tonne)	39
k_Q	$\dfrac{M}{10^6\,kcal}$	Wärmekosten im Brennstoff . . . . Cost of heat in the fuel Coût de la chaleur dans le combustible	39
k_{Qh}	$\dfrac{M}{10^6\,kcal}$	Wärmekosten im beheizten Raum . . . . Cost of heat in the heated space Coût de la chaleur dans le local chauffé	39
l	m	Länge der Rohrleitung . . . . Lenght of pipe line Longueur de la tuyauterie	21
l_k	mm	Koks-Korngröße . . . . Size of coke Grosseur du coke	38
n		Luftüberschußzahl . . . . Excess air factor Coefficient d'excès d'air	14
p_l	$\dfrac{mm\,H_2O}{m}$	Druckgefälle (je 1 m Rohrleitung) . . . . 22, 24, 26, 27, 28, 29 Pressure drop per metre-run of pipe Chute de pression (par m de conduite)	
p_D	ata	Dampfdruck . . . . 18, 40 Steam pressure Pression de la vapeur	
q_f	$\dfrac{kcal}{m^2\,h}$	Wärmeleistung (je 1 m² Fläche) . . . . Heat output per square metre of surface per hour Quantité de chaleur émise (par m² de surface et par heure)	7
q_k	$\dfrac{kcal}{m^2\,h}$	Heizflächenbelastung . . . . Loading of heating surface Taux d'émission de la surface de chauffe	10
q_r	$\dfrac{kcal}{m\,h}$	Wärmeabgabe (je 1 m Rohrleitung) . . . . Heat emission per metre-run of pipe line per hour Chaleur émise par m de conduite	32

q_t	$\dfrac{\text{kcal}}{{}^\circ\text{C}\,\text{h}}$	Wärmeleistung (je 1° C Temperaturunterschied) . . . . . 7 Heat output per hour (per 1° C difference of temperature) Taux d'émission (par $^\circ$C d'écart de température et par heure)
q_F	$\dfrac{\text{kcal}}{\text{m}^2\,\text{h}}$	Wärmebedarf (je 1 m^2 Umschließungsfläche) . . . . . . 9 Heat required (per square metre of wall and window area) Quantité de chaleur nécessaire (par m^2 de surface de paroi extérieure totale)
q_G	$\dfrac{1000\,\text{kcal}}{{}^\circ\text{C}\,(24\,\text{h})}$	Wärmeverbrauch (je 1 Gradtag) . . . . 36 Heat consumption per degree Centigrade per 24 hr. (i. e. per degree-day) Consommation de chaleur (par degré et par jour)
q_J	$\dfrac{\text{kcal}}{\text{m}\,\text{h}\,{}^\circ\text{C}}$	Wärmeverlust (je 1 m Rohrlänge) . . . . 20, 21 Heat loss (per metre-run of pipe) Déperdition de chaleur (par m de conduite)
q_V	$\dfrac{\text{kcal}}{\text{m}^3\,\text{h}}$	Wärmebedarf (je 1 m^3 umbauten Raum) . . . . 8 Head required (per cubic metre of enclosed space) Chaleur nécessaire (par m^3 de local)
r	$\dfrac{\text{m}\,\text{h}\,{}^\circ\text{C}}{\text{kcal}}$	spezifischer Wärmewiderstand . . . . 1, 6 Specific thermal resistance Résistance thermique spécifique
t_a	$^\circ$C	Außentemperatur . . . . 13, 35 Outdoor temperature Température extérieure
t_m	$^\circ$C	mittlere Wassertemperatur . . . . 35 Mean temperature of water Température moyenne de l'eau
$t_{\max}$	$^\circ$C	höchste Betriebstemperatur . . . . 11 Maximum operating temperature Température maximum de fonctionnement
t_v	$^\circ$C	Wassertemperatur im Vorlauf . . . . 23, 35 Temperature of water in flow Température de l'eau dans la canalisation d'amenée
t_r	$^\circ$C	Wassertemperatur im Rücklauf . . . . 23, 35 Temperature of water in return Température de l'eau dans la canalisation de retour
t_L	$^\circ$C	Raumtemperatur . . . . 3, 4, 21 Room temperature Température du local
t_O	$^\circ$C	Oberflächentemperatur . . . . 3, 4 Surface temperature Température superficielle
t_R	$^\circ$C	Temperatur der Rauchgase . . . . 13, 14, 15 Temperature of flue gases Température des fumées

13

G_E $\dfrac{kg}{m}$ Eisengewicht (je 1 m Rohrlänge)   16
Weight of iron (per metre-run of pipe)
Poids de fer (par m de tuyau)

H_o $\dfrac{kcal}{kg}$ oberer Heizwert   57
Gross calorific value
Pouvoir calorifique supérieur

H_u $\dfrac{kcal}{kg}$ unterer Heizwert 14, 36, 37, 39
Net calorific value
Pouvoir calorifique inférieur

J_D $\dfrac{1000\,kcal}{h}$ stündlicher Wärmeinhalt des strömenden Dampfes   18
Heat content of steam flow, per hour
Quantité de chaleur horaire emportée par la vapeur

J_W $\dfrac{1000\,kcal}{h}$ stündlicher Wärmeinhalt des strömenden Wassers   18
Heat content of water flow, per hour
Quantité de chaleur horaire dans l'eau en circulation

M_B $\dfrac{kg}{h}$ stündliche Brennstoffmenge   14
Quantity of fuel per hour
Quantité de combustible par heure

M_W $\dfrac{1}{h}$ stündliche Wassermenge   21
Quantity of water per hour
Quantité d'eau par heure

P_{sch} mm H_2O Schornstein-Zugstärke   13
Chimney draught
Tirage de la cheminée

P_{sch_0} mm H_2O Schornstein-Zugstärke (für 0° C Außentemperatur)   13
Chimney draught (with outdoor temperature 0° C)
Tirage de la cheminée (pour température extérieure de 0° C)

P_{sch_a} mm H_2O Zugstärkenänderung (für andere Außentemperaturen) . . .   13
Variation of draught (for other outdoor temperatures)
Variations du tirage (pour les autres valeurs de la température
extérieure)

ΔP $\dfrac{mm\,H_2O}{m}$ wirksamer Druckunterschied   19
Effective difference of pressure
Différence de pression efficace

Q $\dfrac{kcal}{h}$ Wärmeleistung   7
Heat output
Chaleur fournie par le chauffage

Q_b $\dfrac{1000\,kcal}{h}$ Wärmebedarf des Gebäudes 8, 9, 10
Heat requirements of the building
Quantité de chaleur requise par le bâtiment

Q_h $\dfrac{kcal}{h}$ Wärmeleistung (bezogen auf ein Temperaturgefälle
von 20° C) 22, 23, 26, 28
Heat output (referred to 20° C temperature drop)
Quantité de chaleur émise (pour une chute de température
de 20° C)

Symbol	Unit	Description	Chart
Q_{max}	$\dfrac{kcal}{h}$	höchster Wärmeverbrauch . . . . Maximum heat consumption Consommation maximum de chaleur	36
Q_t	$\dfrac{kcal}{h}$	Wärmeleistung (bei beliebigem Temperaturgefälle) . . . . Heat output (for given temperature drop) Quantité de chaleur émise (pour une chute de température quelconque)	23
Q_D	$\dfrac{1000\,kcal}{h}$	Wärmemenge im niedergeschlagenen Dampf . . . . . . Heat content of condensed steam Chaleur contenue dans la vapeur condensée	30
R	$\dfrac{m^2\,h\,{}^{\circ}C}{kcal}$	Wärmewiderstand . . . . 1, 2, 6, 7 Thermal resistance Résistance thermique	
S		Anzahl der Säulen (des Heizkörpers) . . . . Number of radiator columns Nombre de colonnes (du radiateur)	31
V_b	m^3	umbauter Raum des Gebäudes . . . . Enclosed volume of the building Cube de bâtiment	8
V_g	l	Wasserinhalt der gesamten Heizanlage . . . . Water content of whole heating installation Contenance totale en eau de l'installation	11
V_i	$\dfrac{l}{m^3}$	Rauminhalt (je 1 m Rohrlänge) . . . . Capacity (per metre-run of pipe) Contenance (par m de tuyau)	16
V_z	l	größte Wärmedehnung des Wasserinhalts . . . . Maximum thermal expansion of water-content Dilatation maximum de l'eau contenue dans l'installation	11
V_A	l	notwendiger Rauminhalt des Ausdehnungsgefäßes . . . . . Requisite capacity of expansion tank Capacité nécessaire pour le vase d'expansion	11
V_{R_0}	$\dfrac{Nm^3}{kg}$	Rauminhalt der Rauchgase (für 0° C und 760 mm Hg) . . . Volume of flue gases (at 0° C and 760 mm Hg) Volume spécifique des fumées (ramené à 0° C et 760 mm Hg)	14
W_r	kcal	Wärmeaufwand für einmaliges Hochheizen . . . . . . Heat expenditure for single heating-up Calories à dépenser pour une mise en marche	9
Z_r	mm H_2O	Reibungsverlust . . . . Loss due to friction Perte par frottement	15
Z_w	mm H_2O	Geschwindigkeitsverlust . . . . Loss due to velocity Perte de charge cinétique	15

15

Z_D mm H$_2$O   Druckabfall durch Einzelwiderstände (Dampf)   25
Pressure drop due to individual resistances (steam)
Chute de pression due aux résistances locales (vapeur)

Z_W mm H$_2$O   Druckabfall durch Einzelwiderstände (Warmwasser) . . .   25
Pressure drop due to individual resistances (warm water)
Chute de pression due aux résistances locales (eau chaude)

α_a $\dfrac{\text{kcal}}{\text{m}^2\,\text{h}\,^\circ\text{C}}$ äußere Wärmeübergangszahl 4, 5
Coefficient of outward heat transmission
Coefficient de transmission extérieur

α_i $\dfrac{\text{kcal}}{\text{m}^2\,\text{h}\,^\circ\text{C}}$ innere Wärmeübergangszahl 4, 5
Coefficient of inward heat transmission
Coefficient de transmission intérieur

α_{La} $\dfrac{\text{kcal}}{\text{m}^2\,\text{h}\,^\circ\text{C}}$ äußere Wärmeübergangszahl durch Leitung   3
Coefficient of outward heat transmission, for conduction
Coefficient de transmission extérieur par conductibilité

α_{Li} $\dfrac{\text{kcal}}{\text{m}^2\,\text{h}\,^\circ\text{C}}$ innere Wärmeübergangszahl durch Leitung.   3
Coefficient of inward heat transmission, for conduction
Coefficient intérieur de transmission par conductibilité

α_s $\dfrac{\text{kcal}}{\text{m}^2\,\text{h}\,^\circ\text{C}}$ Wärmeübergangszahl durch Strahlung.   4
Coefficient of heat transmission for radiation
Coefficient de transmission par rayonnement

γ_B $\dfrac{\text{m}^3}{\text{kg}}$ spezifisches Gewicht des Brennstoffs   37
Density of fuel
Poids spécifique de combustible

γ_D $\dfrac{\text{m}^3}{\text{kg}}$ spezifisches Gewicht des Dampfes.   18
Density of steam
Poids spécifique de la vapeur

γ_{L_0} $\dfrac{\text{kg}}{\text{Nm}^3}$ spezifisches Gewicht der Luft (für 0° C und 760 mm Hg) . .   13
Density of air (at 0° C and 760 mm Hg)
Poids spécifique de l'air (ramené à 0° C et 760 mm Hg)

γ_{R_0} $\dfrac{\text{kg}}{\text{Nm}^3}$ spezifisches Gewicht der Rauchgase (für 0° C und 760 mm Hg) 13, 15
Density of flue gases (at 0° C and 760 mm Hg)
Poids spécifique des fumées (ramené a 0° C et 760 mm Hg)

γ_W $\dfrac{\text{kg}}{\text{m}^3}$ spezifisches Gewicht des Wassers   40
Density of water
Poids spécifique de l'eau

δ cm   Schichtdicke (Wandstärke) 1, 2, 3, 6
Thickness of layer (wall thickness)
Epaisseur de couche (Epaisseur de paroi)

δ_J cm   Stärke der Isolierung 2, 20
Thickness of insulation
Epaisseur du revêtement calorifuge

$\varepsilon = \dfrac{1}{\eta_h}$ Heizkennziffer 36, 39

17

Wärmewiderstand und Wärmedurchlässigkeit von Baustoffen.
Thermal Resistance and Heat Permeability of Constructional Materials.
Résistance thermique et perméabilité thermique des matériaux de construction.

δ	cm	Schichtdicke Stoffart	thickness of layer material	épaisseur de couche matière	51 ⑭
r	$\dfrac{m^2\,h\,{}^0C}{kcal}$	spez. Wärmewiderstand	specific thermal resistance	résistance thermique spécifique	1,33
λ	$\dfrac{kcal}{m^2\,h\,{}^0C}$	Wärmeleitzahl	thermal conductivity	conductibilité thermique	0,75
R	$\dfrac{m^2\,h\,{}^0C}{kcal}$	Wärmewiderstand	thermal resistance	résistance thermique	0,68
Λ	$\dfrac{kcal}{m^2\,h\,{}^0C}$	Wärmedurchlässigkeit	heat permeability	perméabilité thermique	1,47

Stoffarten. — Materials. — Genres de matériaux.

				$\lambda =$
⑦	Asbestschiefer	asbestos slate	fibro-ciment	0,19
⑲	Eisenbeton	reinforced concrete	béton armé	1,3
⑰	Kiesbeton ($\gamma = 2200$ kg/m³)	gravel concrete ($\gamma = 2200$ kg/m³)	béton de cailloux ($\gamma = 2200$ kg/m³)	1,1
⑫	Schlackenbetonstein-Mauerwerk	masonry of slag-concrete blocks	maçonnerie en agglomérés de béton de mâchefer	0,6
⑩	Bimsbetonstein-Mauerwerk	masonry of pumice-concrete blocks	maçonnerie en agglomérés de béton-ponce (béton cellulaire)	0,45
⑯	Fliesen und Kacheln	flags and tiles	carreaux et carrelages	0,9
⑨	lufttrockener Gips	air-dry plaster of Paris	plâtre séché naturellement	0,37
⑧	Gipsdielen	plaster tiles	carreaux de plâtre	0,25
⑬	Glas	glass	verre	0,65
⑥	Holz, außen	wood, outside	bois, extérieurement	0,18
④	Holz, innen	wood, inside	bois, intérieurement	0,12
①	Korksteinplatten ($\gamma < 250$ kg/m³)	cork board plates ($\gamma < 250$ kg/m³)	carreaux de liège ($\gamma < 250$ kg/m³)	0,04
②	Korksteinplatten ($\gamma = 250 \div 400$ kg/m³)	cork board plates ($\gamma = 250 \div 400$ kg/m³)	carreaux de liège ($\gamma = 250 \div 400$ kg/m³)	0,055
④	Tekton, Heraklit, gebrannt. Kieselgursteine u. ä.	Tekton, Heraklit, burnt kieselguhr bricks	Tekton, Héraclite, agglomérés de kieselguhr	0,12

Hiezu Fortsetzung auf der Tafelrückseite.

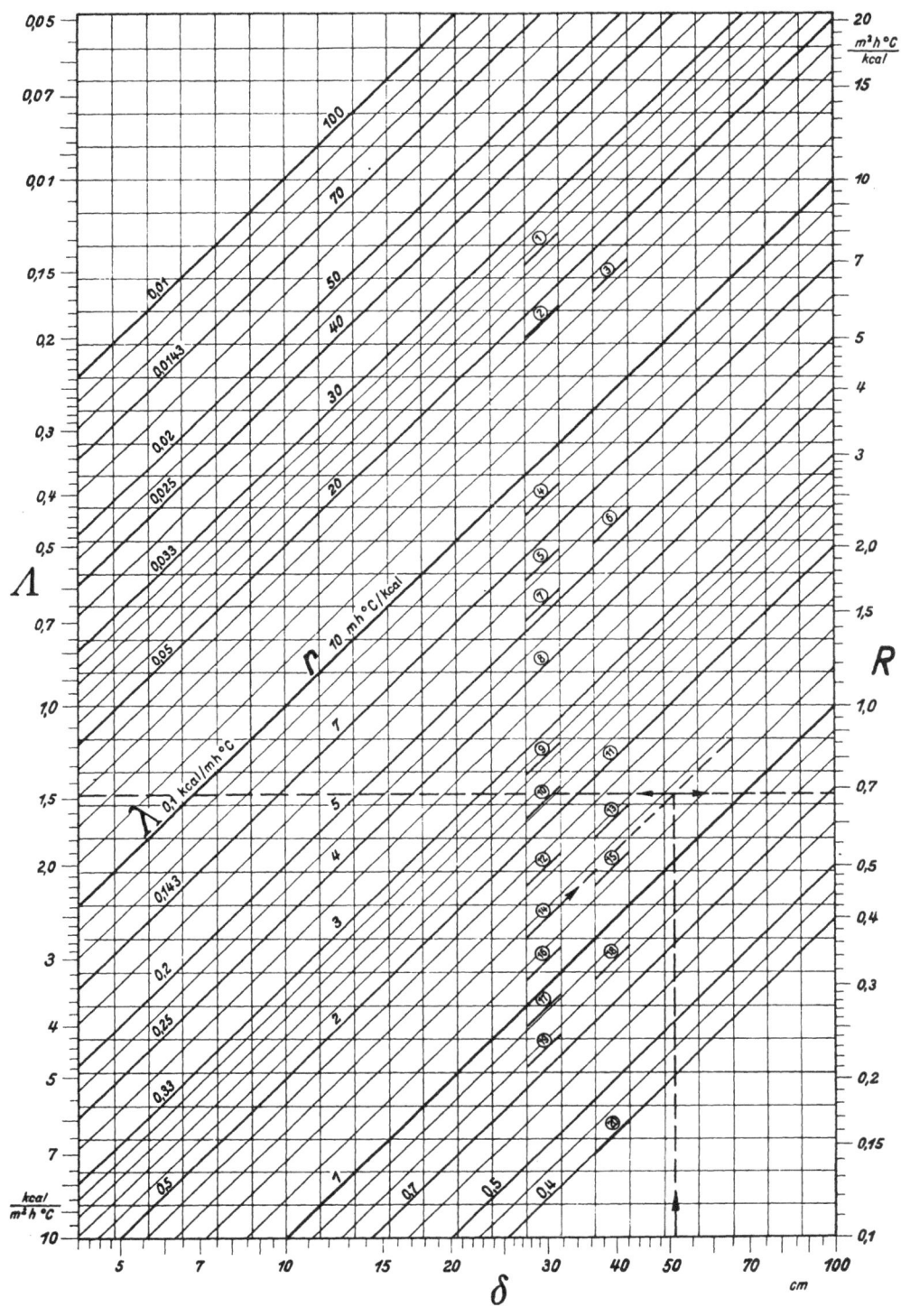

$$\lambda =$$

②	Torfplatten ($\gamma = 250 \div 400$ kg/m³)	peat slabs ($\gamma = 250 \div 400$ kg/m³)	carreaux agglomérés de tourbe ($\gamma = 250 \div 400$ kg/m³)	0,055
①	Torfleichtplatten, kernimprägniert ($\gamma < 250$ kg/m³)	core-impregnated light peat slabs ($\gamma < 250$ kg/m³)	plaques agglomérés légères en tourbe imprégnée à cœur ($\gamma < 250$ kg/m³)	0,04
⑯	Kalksandstein	sand-lime bricks	grès calcaire	0,9
⑳	Granit, Basalt, Gneis, Marmor	granit, gneiss, basalt, marble	granit, gneiss, basalte, marbre	2,5
⑮	Lehm, gestampft	rammed clay	argile pilonnée	0,8
⑥	Linoleum, als Fußbodenbelag	linoleum, as floor covering	linoléum sur parquets	0,16
④	Dachpappe	roofing felt	carton bitumé	0,12
③	Pappe als Wandbelag	millboard used as wall covering	carton employé comme revêtement de murs	0,06
⑭	Kalkputz an Außenflächen	lime plaster on outside surfaces	enduit à la chaux sur surfaces extérieures	0,75
⑫	Kalkputz an Innenflächen	lime plaster on inside surfaces	enduit à la chaux sur surfaces intérieures	0,6
⑪	trockene Sandschüttung in Decken	dry sand filling in ceilings	sable sec (employé comme remplissage de plafond)	0,5
⑱	Schiefer	slate	ardoise	
⑥	Schlackenschüttung in Decken	slag filling in ceilings, etc.	mâchefer (employé comme remplissage de plafonds etc.)	1,2
⑮	Zement, abgebunden	cement, set	ciment (après prise complète)	0,16
⑭	Ziegelstein-Mauerwerk, Außenwand	brickwork, outside wall	maçonnerie de briques (murs extérieurs)	0,75
⑫	Ziegelstein-Mauerwerk, Innenwand	brickwork, inside wall	maçonnerie de briques (murs intérieurs)	0,6

$$R = \frac{r \cdot \delta}{100} = \frac{\delta}{100 \cdot \lambda}$$

$$\Lambda = \frac{100 \cdot \lambda}{\delta} = \frac{100}{r \cdot \delta}$$

DIN 4701. — Rietschel.

	dichte Gesteine (Dolomitkalk- stein, Marmor, Granit, Basalt):	dense stones (Dolomitic lime- stone, marble, granit, basalt):	pierres compactes (calcaire dolomitique, marbre, granit, ba- salte):	
⑲	einseitig, außen	one side, outside	une face, extérieure	
⑱	beiderseits, außen	both sides, outside	deux faces, extérieure	
⑰	beiderseits, innen	both sides, inside	deux faces, intérieure	
	Kiesbeton:	gravel concrete:	béton de cailloux:	
㉕	unverputzt, außen	without plaster, outside	sans enduit, extérieure	
㉔	unverputzt, innen	without plaster, inside	sans enduit, intérieure	
㉓	beiderseits, außen	both sides, outside	deux faces, extérieure	
㉒	beiderseits, innen	both sides, inside	deux faces, intérieure	
	Isolierwände aus Ziegelstein- mauerwerk:	insulating walls of brick ma- sonry	cloisons isolantes en maçonnerie de briques	
⑧	beiderseits ver- putzt, mit Luft- schicht von 5—12 cm	plastered on both sides, with 5 to 12 cm air space	avec enduit sur les deux faces, avec couche d'air de 5 a 12 cm d'épaisseur	
	mit unter Putz ver- legter Isolierung aus kork- oder kernimprägnier- ten Torfleicht- platten an der Innenseite mit einer	with insulation of cork or core-im- pregnated light peat slabs, laid under plaster on inside with a	avec revêtement isolant en carreaux de liège ou en carreaux de tourbe im- prégnés, posé sous en- duit à l'intérieur des locaux, et d'une	
⑤	Stärke	thickness	épaisseur	$\delta_J = 2$ cm
④				3 cm
③				4 cm
②				5 cm
①				10 cm

Wärmewiderstand und Wärmeübergangszahl von Mauerwerk.
Thermal Resistance and Heat Transmission Factor of Masonry.
Résistance thermique et coefficient de transmission de chaleur des maçonneries.

δ	cm	Wandstärke	wall thickness	épaisseur de paroi	51
		Bauart des Mauerwerks	type of masonry	genre de maçonnerie	⑪
R	$\dfrac{m^2 \, h \, {}^0C}{kcal}$	Wärmewiderstand	thermal resistance	résistance thermique	0,9
k	$\dfrac{kcal}{m^2 \, h \, {}^0C}$	Wärmedurchgangszahl	coefficient of heat transmission	coefficient de transmission thermique	1,11

Bauarten des Mauerwerks — Types of Masonry — Genres de maçonneries

	Ziegelsteine:	bricks:	brique:
⑪	einseitig verputzt, Außenwand	plastered on one side, outside wall	avec enduit sur une face, paroi extérieure
⑩	beiderseitig verputzt, Außenwand	plastered on both sides, outside wall	avec enduit sur les deux faces, paroi extérieure
⑥	beiderseitig verputzt, Innenwand	plastered on both sides, inside wall	avec enduit sur les deux faces, paroi intérieure
	Schlackenbetonsteine:	slag concrete blocks:	agglomérés de béton de mâchefer:
⑦	beiderseits, außen	both sides, outside	deux faces, extérieure
⑥	beiderseits, innen	both sides, inside	deux faces, intérieure
	Bimsbetonsteine, Schwemmsteine:	pumice concrete blocks, porous bricks:	agglomérés de béton-ponce, béton cellulaire:
㉑	beiderseits, außen	both sides, outside	deux faces, extérieure
⑳	beiderseits, innen	both sides, inside	deux faces, intérieure
	Kalksandsteine:	sand-lime bricks:	grès calcaire:
⑬	einseitig, außen	one side, outside	une face, extérieure
⑫	beiderseits, außen	both sides, outside	deux faces, extérieure
⑨	beiderseits, innen	both sides, inside	deux faces, intérieure
	porige Gesteine (Sandstein, weicher oder sandiger Kalkstein):	porous stones (sandstone, soft or sandy limestone):	pierres poreuses (grès, calcaire tendre ou sableux):
⑯	einseitig, außen	one side, outside	une face, extérieure
⑮	beiderseits, außen	both sides, outside	deux faces, extérieure
⑭	beiderseits, innen	both sides, inside	deux faces, intérieure

Hiezu Fortsetzung auf der Vorderseite des Blattes.

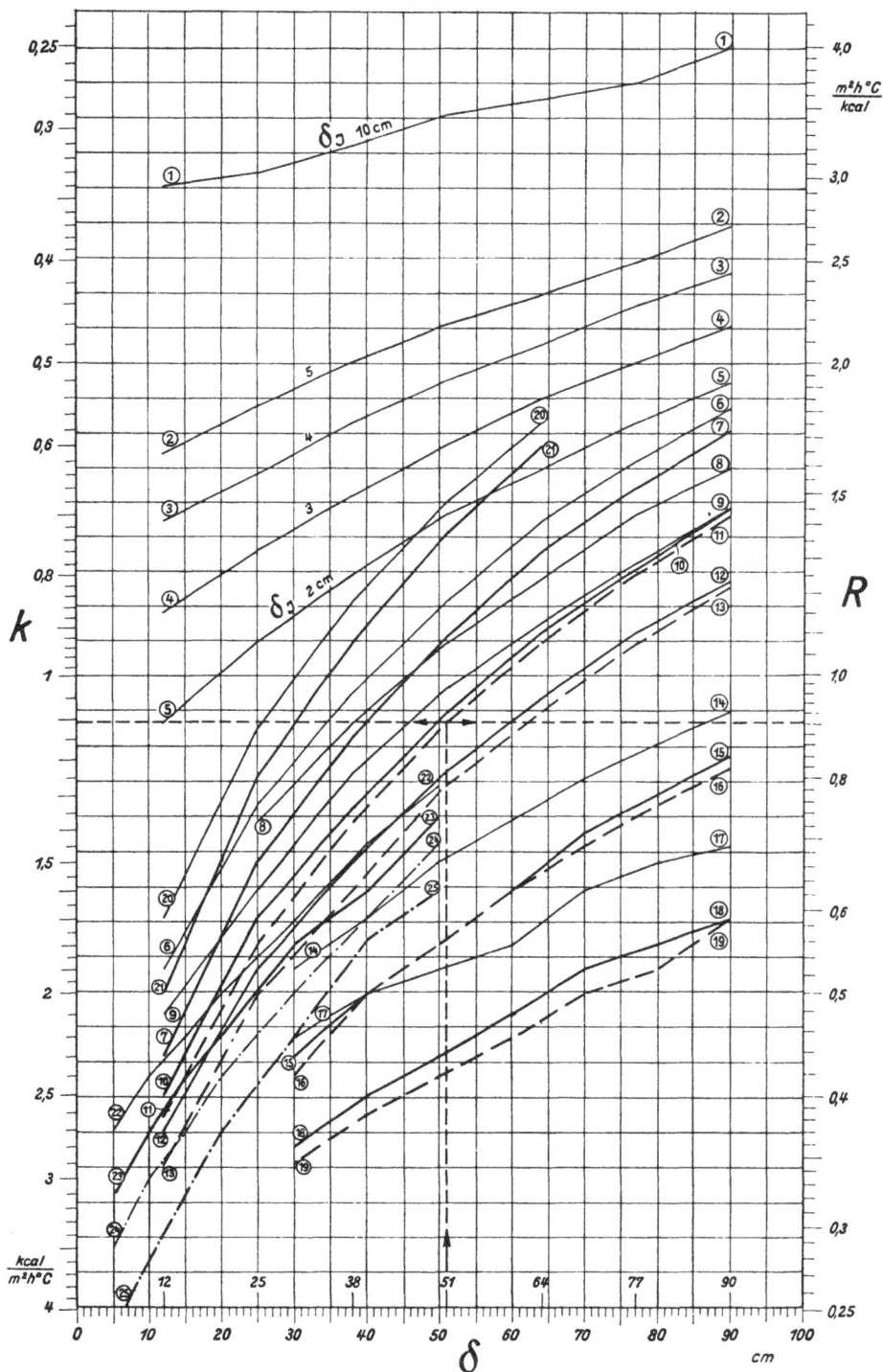

3

Wärmeübergang durch Leitung.
Heat Transmission by Conduction.
Transmission de chaleur par conduction.

I. Innerer Wärmeübergang. — Inward Heat Transmission. — Transmission de chaleur vers l'intérieur.

δ	cm	Wandstärke	wall thickness	épaisseur de paroi	51
		Wandart	nature of wall	nature de la paroi	MW
t_L	°C	Raumtemperatur	room temperature	température du local	20,0
t_O	°C	Oberflächentemperatur (innen)	surface temperature (inside)	température superficielle (intérieure)	15,0
$t_L - t_O$	°C	wirksamer Temperaturunterschied	effective temperature difference	différence de température efficace	5,0
		Luftströmung im Raum	air flow in room	circulation de l'air dans le local	②
α_{Li}	$\dfrac{\text{kcal}}{\text{m}^2\,\text{h}\,°\text{C}}$	innere Wärmeübergangszahl durch Leitung	coefficient of inward heat transmission by conduction	coefficient de transmission intérieure par conductibilité	3,3

Wandarten. — Nature of Walls. — Nature des parois.

MW	Mauerwerk	brickwork	maçonnerie
EF	Einfachfenster	single-glass window	fenêtres simples
	Doppelfenster	double-glass window	doubles fenêtres

Luftströmung im Raum. — Air Flow in Room. — Circulation de l'air dans le local.

①	störungsfrei strömende Luft	undisturbed air flow	circulation sans perturbation	3
②	normale Raumluft	ordinary room conditions	atmosphère normale	4
③	Raumluft bei großen Temperaturunterschieden (Einzelfenster)	room with large temperature difference (with single-glass window)	grands écarts de températures (fenêtres simples)	5
④	gestörte Raumluft (Eisenbahnfenster)	disturbed air (in a railway carriage)	atmosphère agitée (compartiment de chemin de fer)	6

$$\alpha_{Li} = 0{,}55 \cdot m\,[t_L - t_O]^{0,25}$$

$$\alpha_i = \alpha_{Li} + \alpha_{Si}$$

II. Äußerer Wärmeübergang. — Outward Heat Transmission. — Transmission de chaleur vers l'extérieur.

w_L	$\dfrac{\text{m}}{\text{s}}$	Windgeschwindigkeit	air velocity	vitesse de l'air	6,0
α_{La}	$\dfrac{\text{kcal}}{\text{m}^2\,\text{h}\,°\text{C}}$	äußere Wärmeübergangszahl durch Leitung	coefficient of outward heat transmission by conduction	coefficient de transmission extérieure par conductibilité	26,5

$$\alpha_{La} = 5{,}3 + 3{,}6 \cdot w_L \quad \ldots \ldots \quad (w_L \leqq 5\ \text{m/s})$$

$$\alpha_{La} = 6{,}47 \cdot w_L \quad \ldots \ldots \quad (w_L > 5\ \text{m/s})$$

$$\alpha_a = \alpha_{La} + \alpha_{Sa}$$

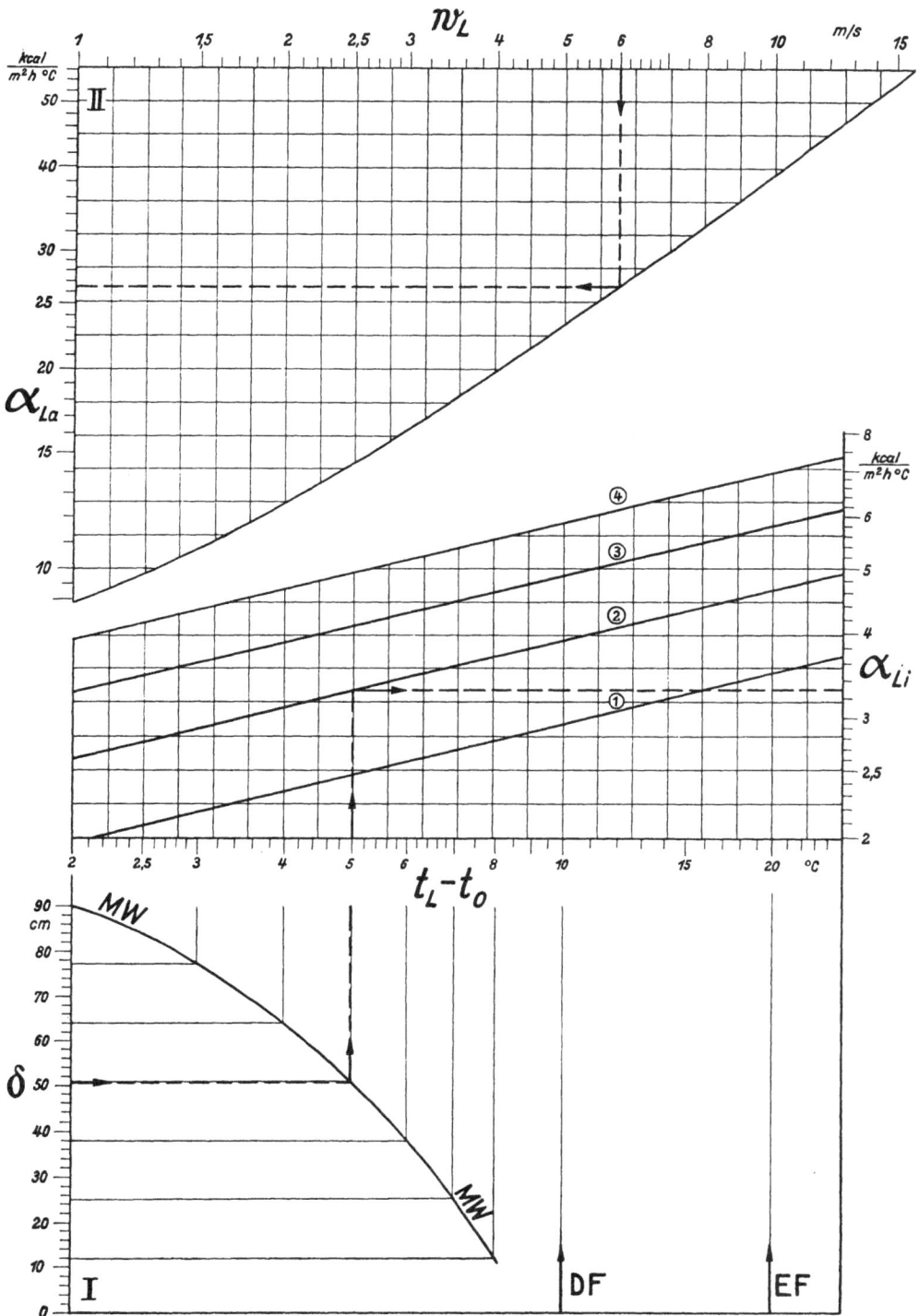

4

Wärmeübergang durch Strahlung.
Heat Transmission by Radiation.
Transmission de chaleur par rayonnement.

I.

t_o	°C	Oberflächentempe-ratur	surface tempera-ture	température super-ficielle	15,0
t_L	°C	Raumtemperatur	room temperature	température du local	20,0
C_S	$\dfrac{\text{kcal}}{\text{m}^2\,\text{h}\,(^o\text{abs})^4}$	Strahlungskon-stante	radiation constant	constante de rayonnement	4,6
α_S	$\dfrac{\text{kcal}}{\text{m}^2\,\text{h}\,^o\text{C}}$	Wärmeübergangs-zahl durch Strah-lung	coefficient of heat transmission by radiation	coefficient de trans-mission par rayonnement	4,55

$$\alpha_S = C_S\,\frac{\left(\dfrac{T_L}{100}\right)^4 - \left(\dfrac{T_o}{100}\right)^4}{t_L - t_o}$$

II.

		Stoffart	Material	Matérial	③
C_S	$\dfrac{\text{kcal}}{\text{m}^2\,\text{h}\,(^o\text{abs})^4}$	Strahlungskon-stante	radiation constant	constante de rayon-nement	4,6
$\dfrac{C_S}{4,96}$		Absorptionsver-hältnis	absorption ratio	taux d'absorption	0,93

Stoffarten. — Materials. — Matériaux.

	Stoffart	Material	Matérial	$C_S =$
①	Eisen, matt oxy-diert	iron, matte oxidis-ed	fer, oxydé mat	4,76
㉑	Kupfer, blank poliert	copper, polished bright	cuivre poli	0,85
⑭	Kupfer, gewalzt	copper, rolled	cuivre laminé	3,17
⑳	Zink, matt	zinc, matte	zinc, mat	1,04
⑧	Holz, glatt	wood, smooth	bois, lisse	1,86
⑮	Marmor ⎱ glatt ge-schliffen,	marble ⎱ ground smooth,	marbre⎱ surfaces doucies lisses,	2,88
⑯	Granit ⎰ aber nicht glänzend	granite ⎰ but not polished	granit ⎰ mais non brillantes	2,33
⑦	Gips	plaster of Paris	plâtre	3,86
⑤	Kalkmörtel, rauh weiß	lime mortar, rough white	mortier de chaux, blanc rugueux	4,47
②	Verputz	plaster	enduits de murs in-térieurs	4,61
③	Mauerwerk	masonry	maçonnerie	4,61
⑲	Kies	gravel	gravier	1,44
⑰	Lehm	clay	argile	1,93
⑪	Sand	sand	sable	3,77
④	Glas	glass	verre	4,61
⑬	Wasser	water	eau	3,32
⑫	Sägespäne	sawdust	sciure de bois	3,72
⑥	Papier	paper	papier	3,96
⑱	Ackererde	soil	terre végétale	1,89
⑩	Baumwollzeug	cotton fabric	tissu de coton	3,82
⑨	Ölanstrich	oil paint	peinture à l'huile	3,86

Hütte.

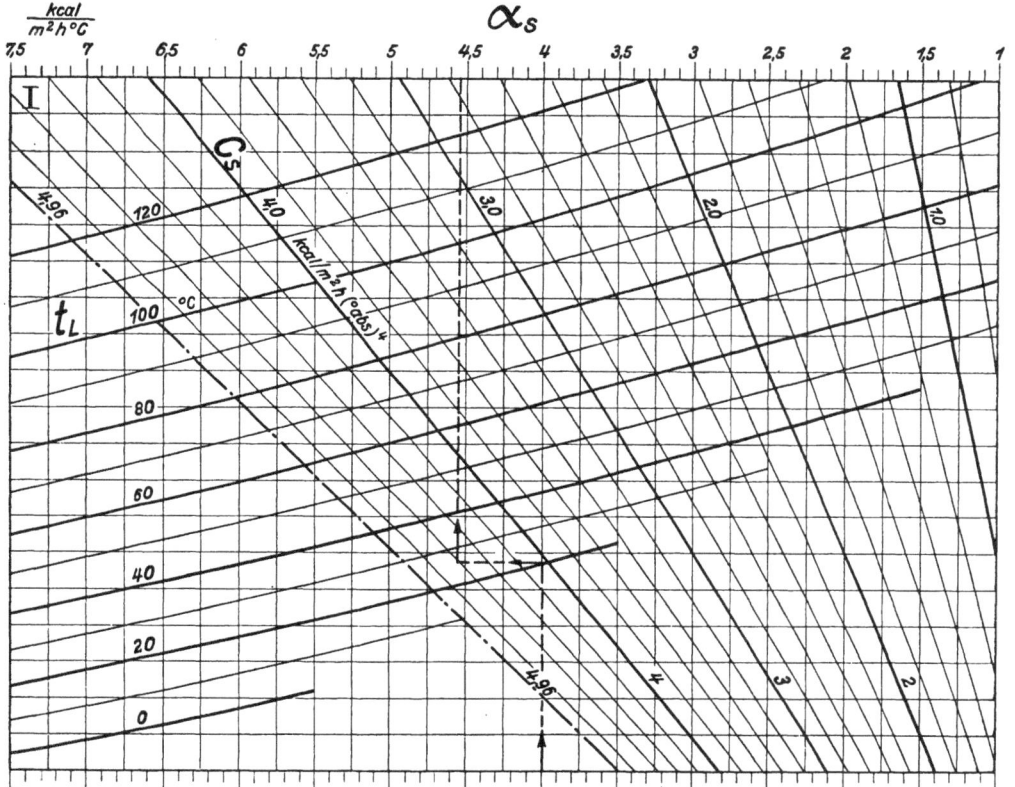

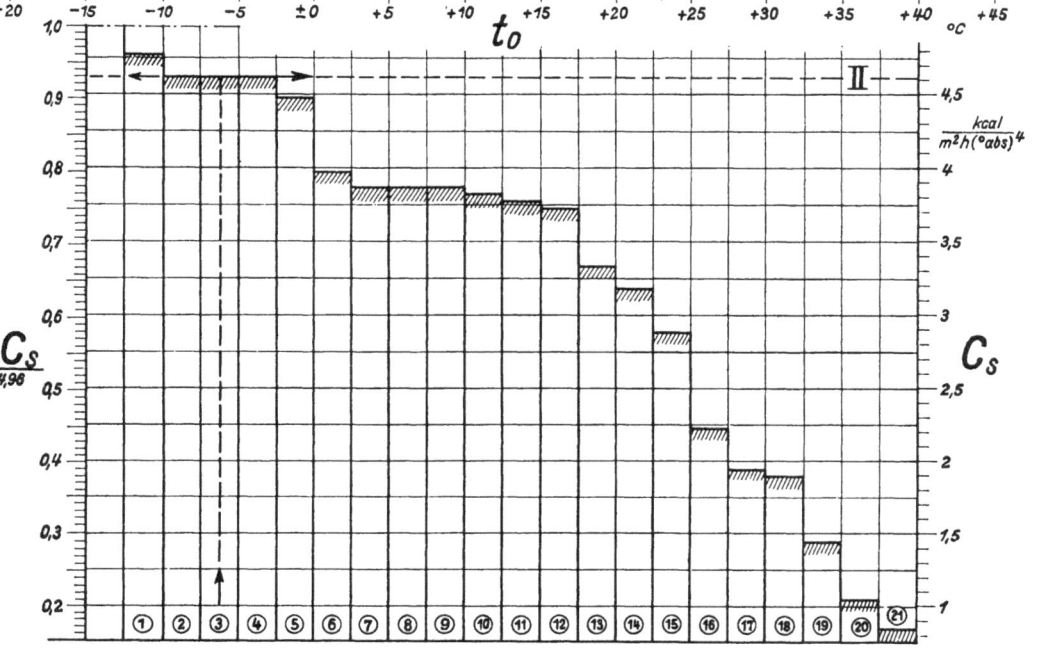

5

Berechnung der Wärmedurchgangszahl.
Calculation of the Coefficient of Heat Transmission.
Calcul du coefficient de transmission de chaleur.

①

α_i	$\dfrac{kcal}{m^2\,h\,{}^{0}C}$	innere Wärme-durchgangszahl	coefficient of inward heat transmission	coefficient de trans-mission intérieur	7,0
Λ	$\dfrac{kcal}{m^2\,h\,{}^{0}C}$	Wärmedurchlässig-keit	heat permeability	perméabilité ther-mique	1,5
k'	$\dfrac{kcal}{m^2\,h\,{}^{0}C}$	Teil-Wärmedurch-gangszahl	coefficient of partial heat transmission	coefficient de trans-mission partiel	1,24

②

α_a	$\dfrac{kcal}{m^2\,h\,{}^{0}C}$	äußere Wärme-durchgangszahl	coefficient of out-ward heat trans-mission	coefficient de trans-mission extérieur	20,0
k'	$\dfrac{kcal}{m^2\,h\,{}^{0}C}$	Teil-Wärmedurch-gangszahl	coefficient of par-tial heat trans-mission	coefficient de trans-mission partiel	1,24
k	$\dfrac{kcal}{m^2\,h\,{}^{0}C}$	Wärmedurch-gangszahl	coefficient of heat transmission	coefficient de trans-mission de cha-leur	1,17

$$k = \cfrac{1}{\dfrac{1}{\alpha_1} + \dfrac{1}{\Lambda} + \dfrac{1}{\alpha_2}}$$

$$\frac{1}{k'} = \frac{1}{\alpha_i} + \frac{1}{\Lambda}$$

$$\frac{1}{k} = \frac{1}{k'} + \frac{1}{\alpha_a}$$

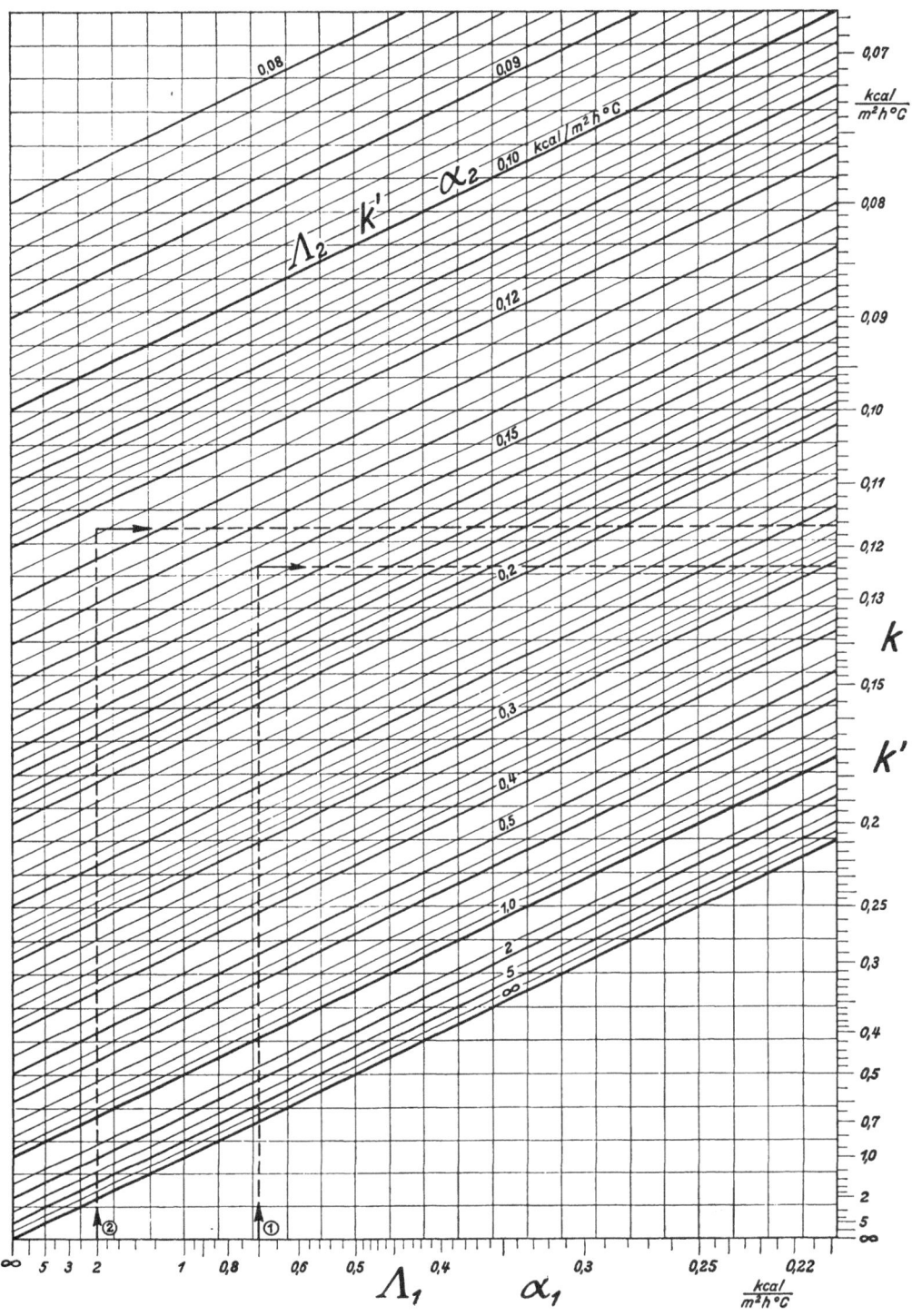

6

Wärmewiderstand und Wärmedurchlässigkeit von Metallen und Luftschichten.
Thermal Resistance and Heat Permeability of Metal and Air Spaces.
Résistance thermique et perméabilité thermique des métaux et des couches d'air.

I. Metalle. — Metals. — Métaux.

δ	cm	Schichtdicke	thickness of layer	épaisseur de couche	2,0
		Metallart	metal	métal	Fe
R	$\dfrac{m^2\,h\,^0C}{kcal}$	Wärmewiderstand	thermal resistance	résistance ther- mique	0,000445
k	$\dfrac{kcal}{m^2\,h\,^0C}$	Wärmedurchlässig- keit	heat permeability	perméabilité ther- mique	2250

Metallarten. — Metals. — Métaux.

Al	Aluminium	aluminium	aluminium
Cu	Kupfer	copper	cuivre
Fe	Eisen	iron	fer
Ni	Nickel	nickel	nickel
Pb	Blei	lead	plomb
Sb	Zinn	tin	étain
Zn	Zink	zinc	zinc

II. Luftschichten. — Air Spaces. — Couches d'air.

δ	cm	Schichtdicke	thickness of layer	épaisseur de couche	4,0
		Lage der Luft- schicht	position of air space	position de la couche d'air	⚏
R	$\dfrac{m^2\,h\,^0C}{kcal}$	Wärmewiderstand	thermal resistance	résistance ther- mique	0,185
Λ	$\dfrac{kcal}{m^2\,h\,^0C}$	Wärmedurchlässig- keit	heat permeability	perméabilité ther- mique	5,4

Lage der Luftschicht. — Position of Air Space. — Position de la couche d'air.

‖	senkrechte Luft- schicht	vertical air space	couche d'air verti- cale
⚏	waagerechte Luft- schicht mit Wärmestrom nach oben	horizontal air space with upward heat flow	couche d'air hori- zontale, flux de chaleur ascendant
⚎	waagerechte Luft- schicht mit Wärmestrom nach unten	horizontal air space with downward heat flow	couche d'air hori- zontale, flux de chaleur descen- dant

DIN 4701. — Hütte. — Rietschel.

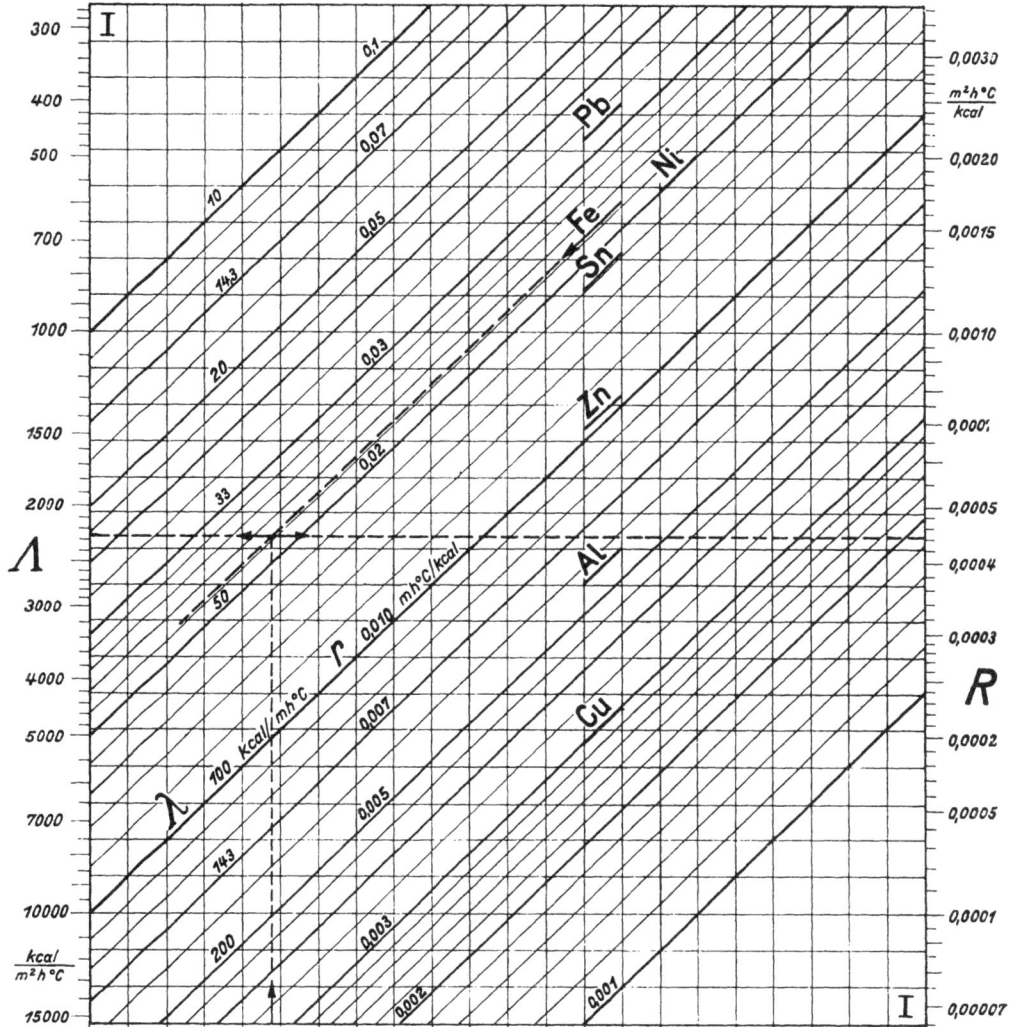

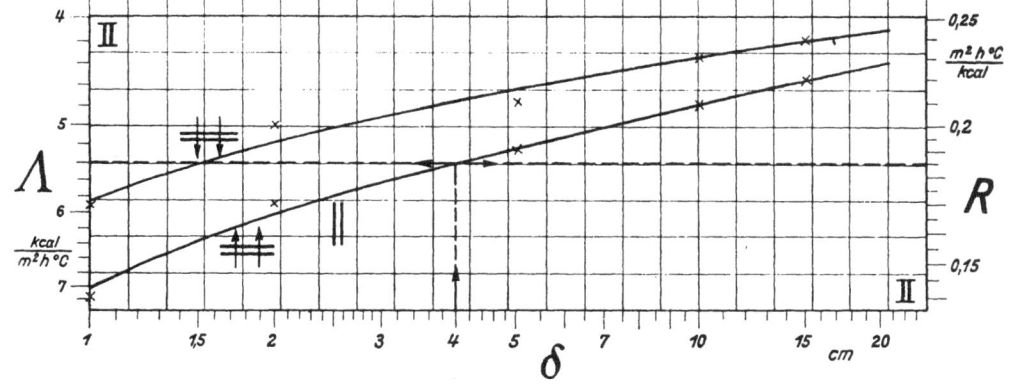

7

Wärmedurchgangszahl und Wärmebedarf.
Coefficient of Heat Transmission and Heat Requirement.
Coefficient de transmission de chaleur et besoins en chaleur.

①

k	$\dfrac{kcal}{m^2\,h\,^0C}$	Wärmedurch-gangszahl	coefficient of heat transmission	coefficient de trans-mission de cha-leur	1,43
F_q	m^2	Durchgangsfläche	transmission sur-face	surface de transmis-sion	24,0
q_t	$\dfrac{kcal}{^0C\,h}$	spez. Wärmelei-stung (je 1^0 Tem-peraturunter-schied)	heat output per hour (per 1^0 C difference tem-perature)	taux de transmis-sion (par ^{0}C d'é-cart de tempéra-ture et par heure)	34,2
$\varDelta t$	^0C	Temperaturunter-schied	temperature differ-ence	différence de tem-pérature	9,0
Q	$\dfrac{kcal}{h}$	Wärmeleistung	heat output	chaleur fournie par le chauffage	310

②

q_f	$\dfrac{kcal}{m^2\,h}$	spez. Wärmelei-stung (je 1 m^3 Durchgangs-fläche)	heat output per square metre of surface per hour	quantité de chaleur transmise (par m^2 de surface et par heure)	12,8

$$q_f = \varDelta t \cdot k = \frac{\varDelta t}{R}$$

$$q_t = F \cdot k = \frac{F}{R}$$

$$Q = F \cdot q_f = \varDelta t \cdot q_t$$

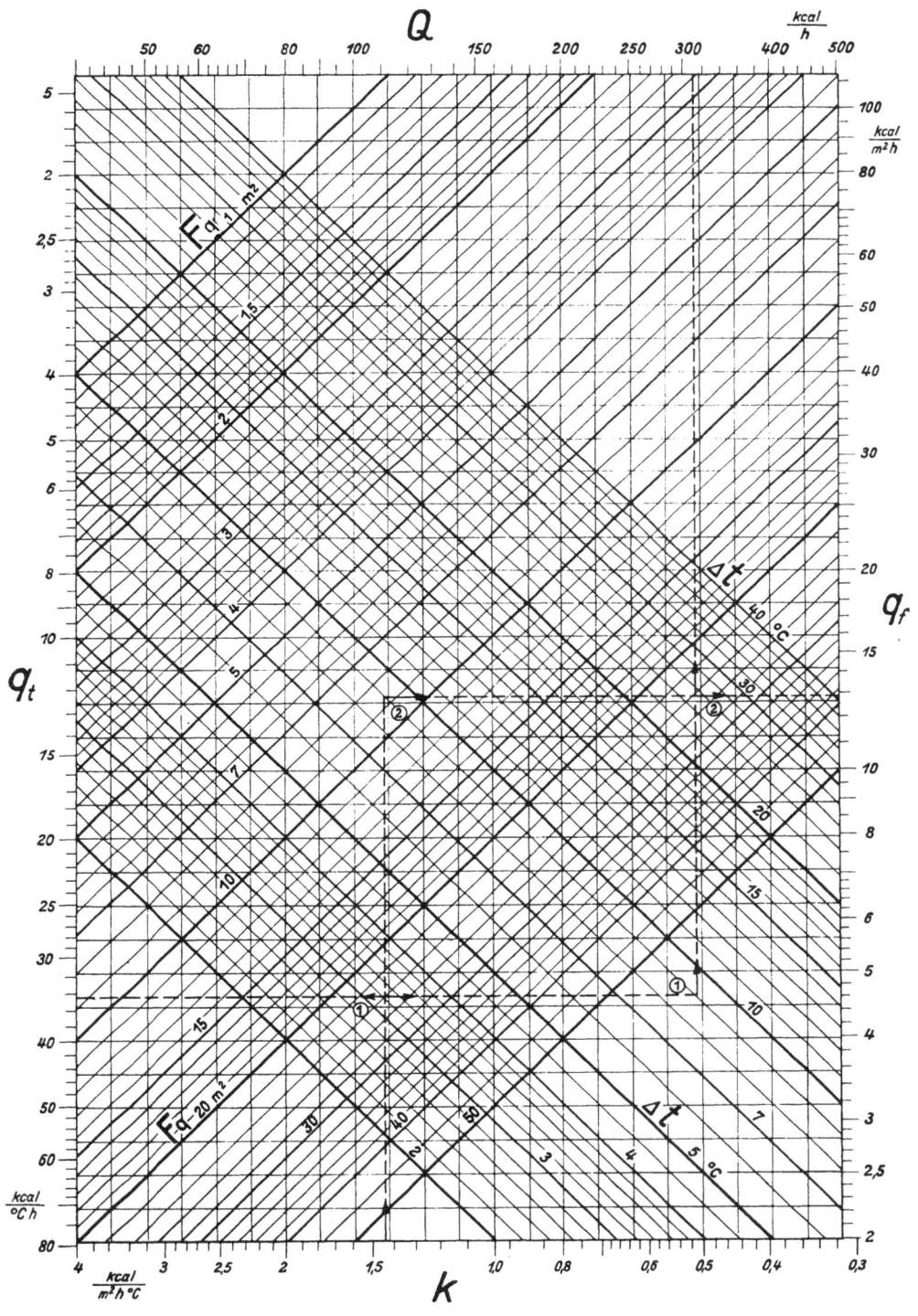

Wärmebedarf von Gebäuden (Mittelwerte).
Heat Requirements of Buildings (Average Values).
Besoins en chaleur des édifices (valeurs moyennes).

					①	②
V_b	m³	umbauter Raum	enclosed volume of the building	cube du bâtiment	6000	6000
		Ausführungsart	type of construction	genre d'exécution	ⒶⒷ	
q_V	$\dfrac{kcal}{m^3\,h}$	Wärmebedarf (ie 1 m³ umbauten Raum)	heat required (per cubic metre of enclosed space)	chaleur nécessaire (par m³ de local)	16,0	26,5
Q_b	$\dfrac{1000\,kcal}{h}$	Wärmebedarf des Gebäudes	heat requirements of the buildings	quantité de chaleur requise par le bâtiment	96	160

Ausführungsarten. — Type of Construction. — Genres d'exécution.

Ⓐ	gute Ausführung, günstige Verhältnisse	good construction favourable conditions	bonne exécution, conditions favorables
Ⓑ	schlechte Ausführung, ungünstige Verhältnisse	bad construction, unfavourable conditions	mauvaise exécution, conditions défavorables

$$\boxed{Q_b = q_V \cdot V_b}$$

Rietschel. — Hottinger M., Heizung und Lüftung. München 1926.

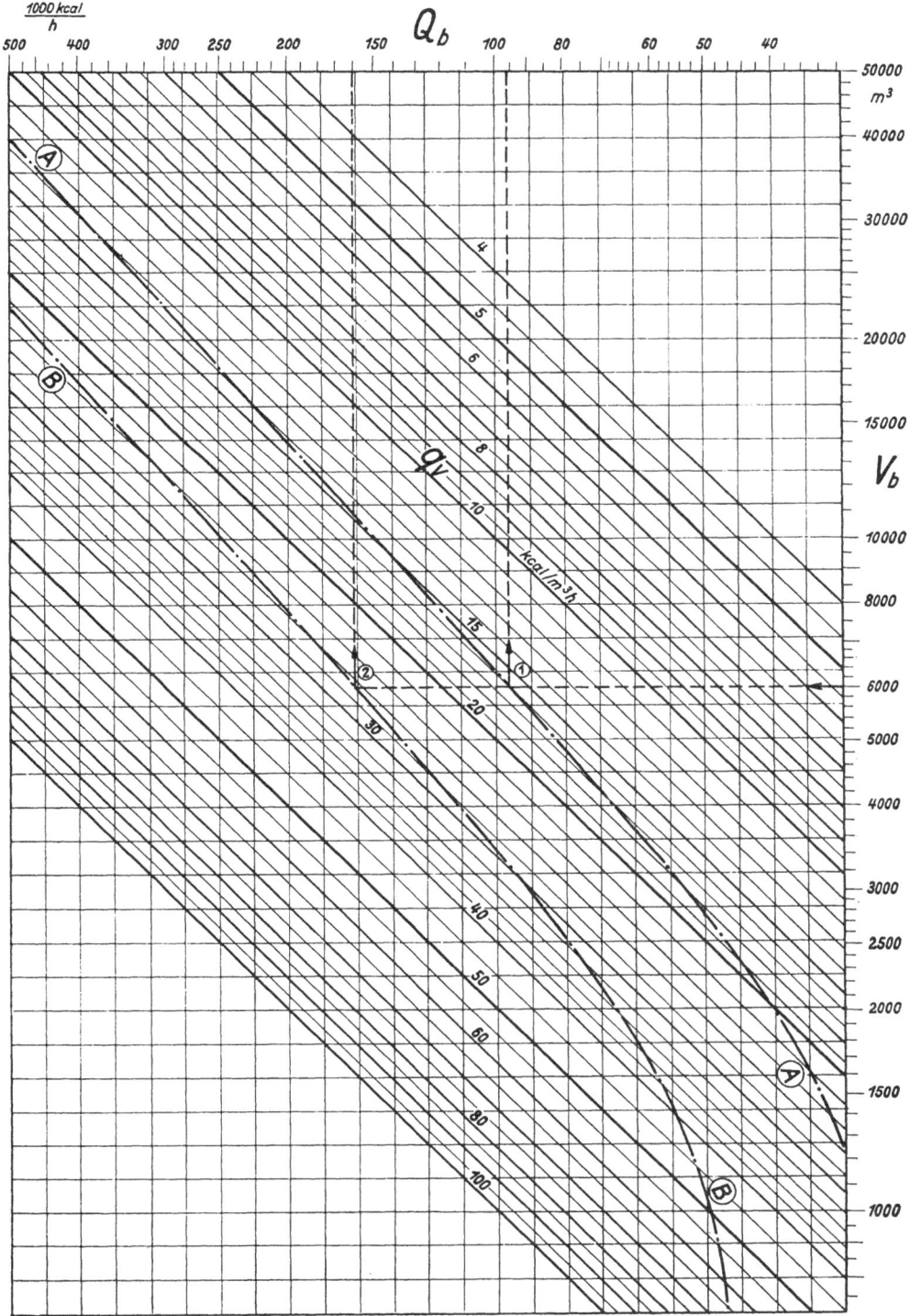

9

Kirchenheizung.
Church Heating.
Chauffage des églises.

Z	h	Aufheizzeit	heating-up period	temps de mise en route	5,0
F_0	m²	Umschließungsfläche	exposed outer surface of building	surface totale de paroi extérieure	3600
F_F	m²	Fensterfläche	window area	surface des fenêtres	900
$\sigma = \dfrac{F_F}{F_0}$		Fensterverhältnis	window ratio	proportion de surface vitrée	0,25
q_F	$\dfrac{\text{kcal}}{\text{m}^2\,\text{h}}$	Wärmebedarf (je 1 m² Umschließungsfläche)	heat required (per square metre of wall and window area)	quantité de chaleur nécessaire (par m² de surface extérieure totale)	77
Q_b	$\dfrac{1000\,\text{kcal}}{\text{h}}$	Wärmebedarf	heat requirement	quantité de chaleur requise	277
w_F	$\dfrac{\text{kcal}}{\text{m}^2}$	Wärmeaufwand (je 1 m² Umschließungsfläche)	heat expenditure (per square metre of wall and window area)	dépense de chaleur (par m² de surface extérieure totale)	355
W_b	1000 kcal	Wärmeaufwand für einmaliges Hochheizen	heat expenditure for single heating-up	calories à dépenser pour une mise en marche	1280
		für:	for:	pour:	
t_L	°C	Raumtemperatur	room temperature	température du local	$+\,12$
t_a	°C	Außentemperatur	outdoor temperature	température extérieure	$-\,15$

Gröber-Sieler, Wärmebedarfsbestimmung von Kirchen. München 1935.

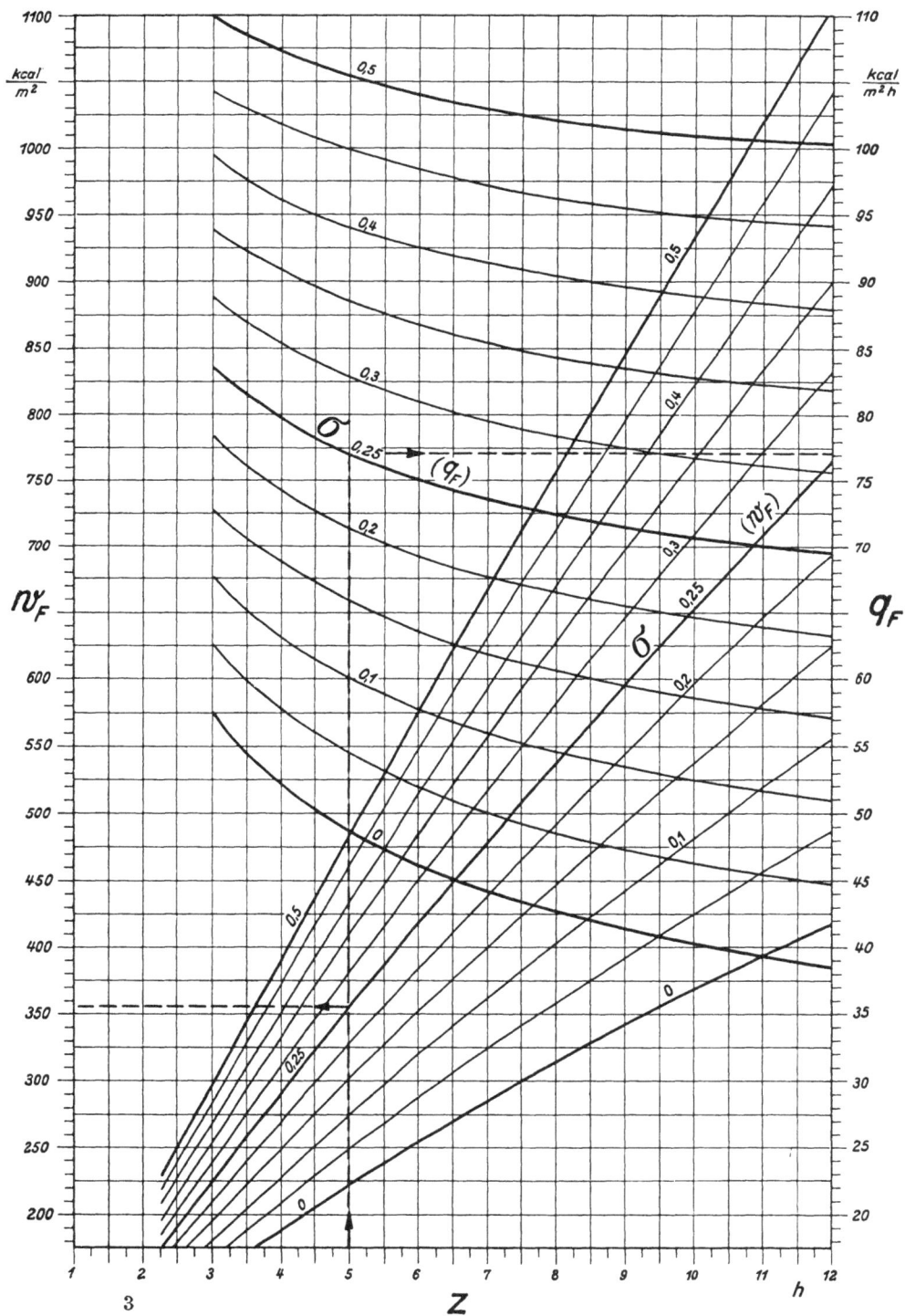

10

Kesselheizfläche.
Boiler Heating Surface.
Surface de chauffe de la chaudière.

Q_b	$\dfrac{1000 \text{ kcal}}{\text{h}}$	Wärmebedarf des Gebäudes	heat requirements of the building	quantité de chaleur requise par le bâtiment	130
	$\%$	Zuschlag für Wärmeverluste	allowance for heat losses	majoration pour pertes de chaleur	10
q_k	$\dfrac{\text{kcal}}{\text{m}^2 \text{ h}}$	Heizflächenbelastung	loading of heating surface	taux d'émission de la surface de chauffe	8000
		Kesselart	boiler type	type de chaudière	mZ—W
		Brennstoffart	kind of fuel	nature du combustible	KK
F_k	m^2	Kesselheizfläche	boiler heating surface	surface de chauffe de la chaudière	17,8

Zuschlag für Wärmeverluste. — Allowance for Heat Losses. — Majoration pour pertes de chaleur.

5%	geschützte Rohrleitung	fully-protected pipe line	tuyauterie protégée
10%	weniger geschützte Rohrleitung	less-protected pipe line	tuyauterie médiocrement protégée
15%	besonders ungünstig liegende Rohrleitung	pipe line with specially unfavourable conditions	tuyauterie placée dans une situation particulièrement défavorable

Kesselarten (Glieder- und schmiedeeiserne Kessel). — Types of Boilers (Sectional and wrought iron boilers). — Types de chaudières (chaudières fonte en sections et chaudières en tôle).

oZ	ohne Züge	without flues	sans carneaux intérieurs
mZ	mit Züge	with flues	avec carneaux intérieurs
W	WarmwasserKessel	hot water boiler	chaudière à eau chaude
D	NiederdruckdampfKessel	low pressure steam boiler	chaudière à vapeur à basse pression

Brennstoffarten. — Kinds of Fuel. — Genres de combustible.

KK	Koks oder Kohle	coke or coal	coke ou houille
BB	Braunkohle oder Braunkohlenbriketts	lignite or lignite briquette	lignite ou briquettes de lignite

DIN 4701. — Recknagel.

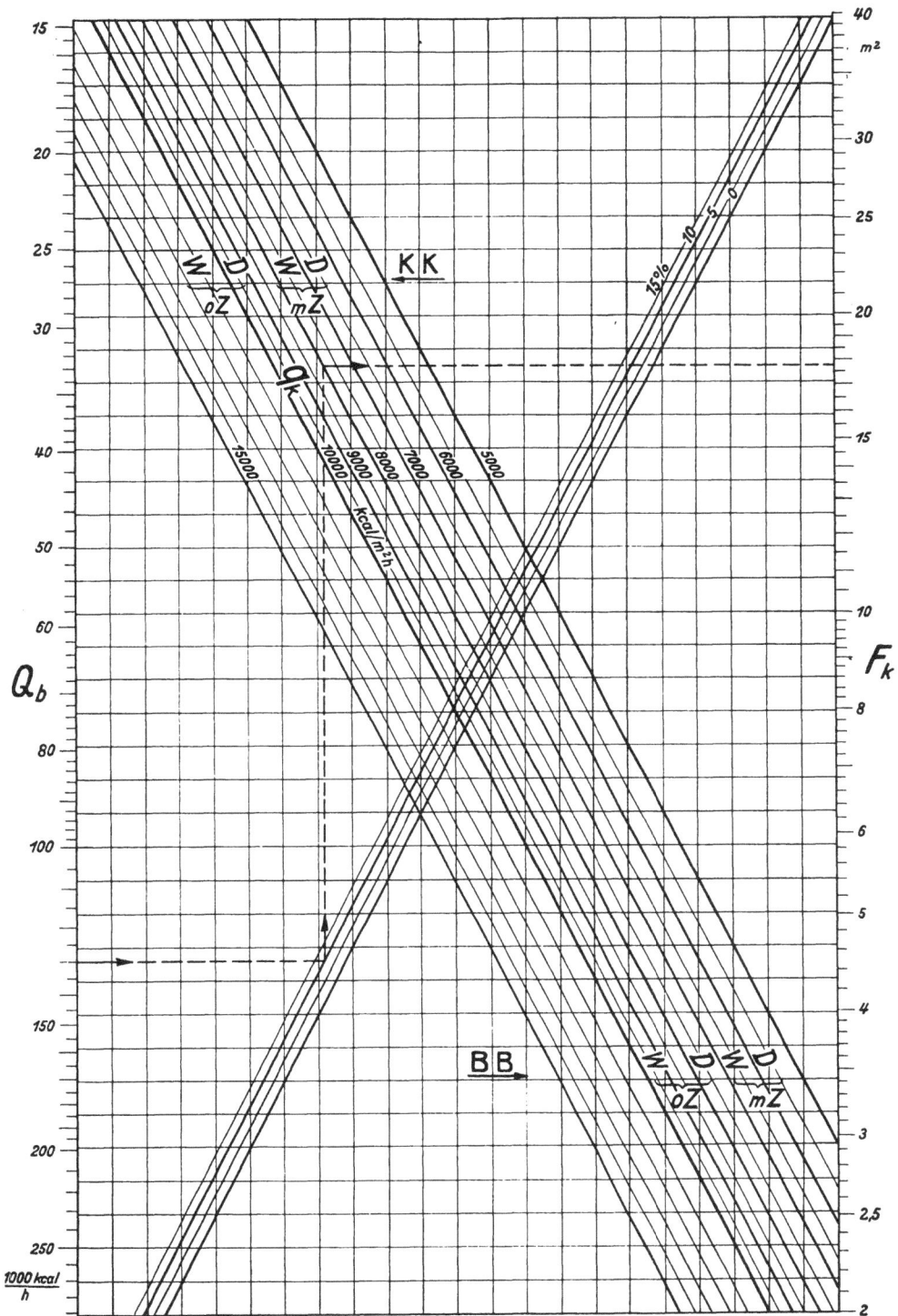

11

Ausdehnungsgefäß.
Expansion Tank.
Vase d'expansion.

t_{max}	°C	höchste Betriebstemperatur (der Warmwasserheizung)	maximum operating temperature (of hot water system)	température maximum de fonctionnement (de chauffage à eau chaude)	95
V_g	l	Wasserinhalt der gesamten Heizanlage	water content of whole heating installation	contenance totale en eau de l'installation	340
V_z	l	größte Wärmedehnung des Wasserinhalts	maximum thermal expansion of water content	dilatation maximum de l'eau contenue dans l'installation	13,5
V_A	l	notwendiger Rauminhalt des Ausdehnungsgefäßes	requisite capacity of expansion tank	capacité nécessaire pour le vase d'expansion	27,0

$$V_z = \frac{v_{max} - 1}{1} \cdot V_g$$

$$V_A = 2 \cdot V_z$$

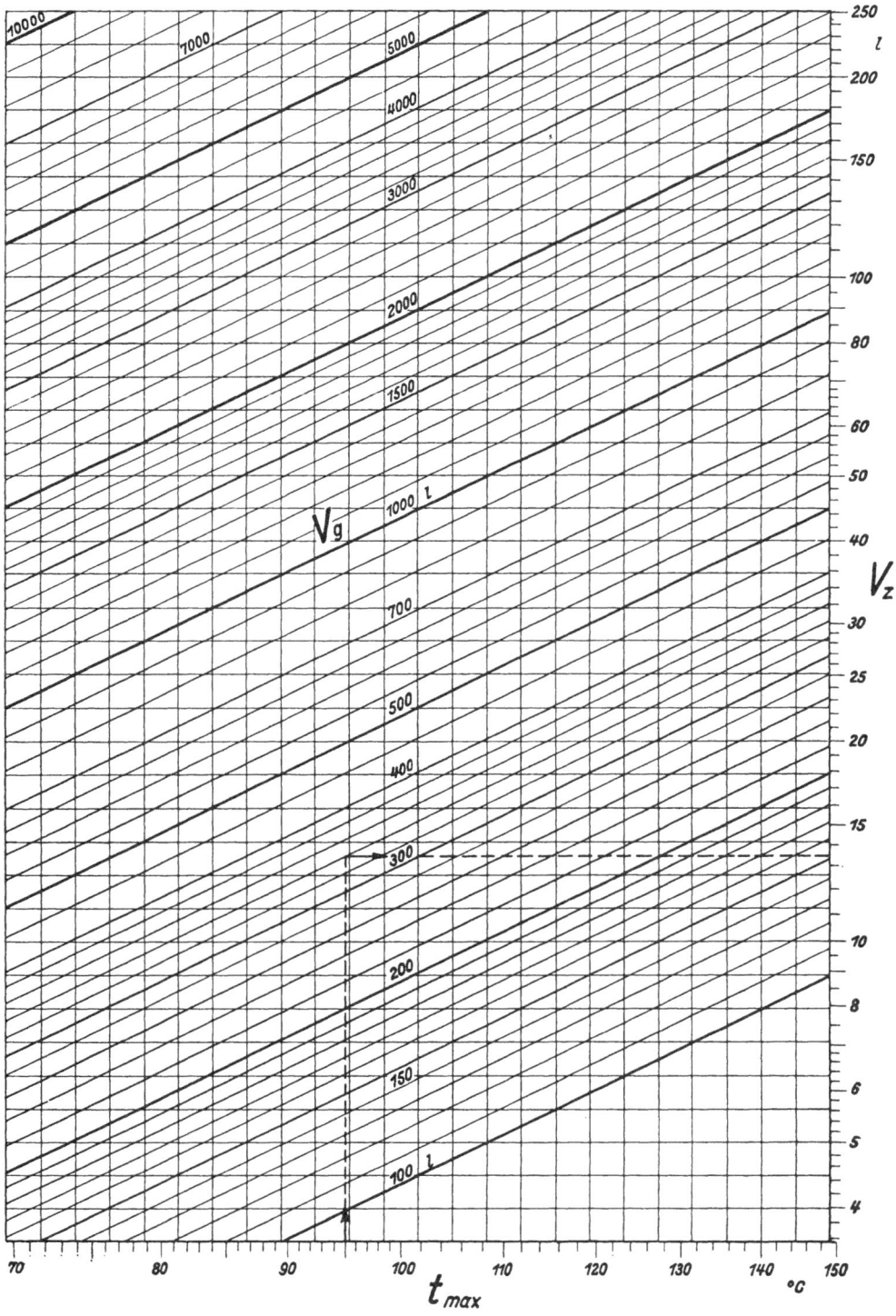

12

Sicherheitsleitungen.
Safety Pipes.
Conduites de sûreté.

W Warmwasserheizung (Sicherheitsleitungen). — Hot Water Heating (Safety Pipes). — Chauffage à eau chaude (conduites de sûreté).

		Ausführungsart	method of construction	genre d'exécution	WA
		Leitungsart	nature of pipe line	genre de conduite	1
F_k	m²	Kesselheizfläche	boiler heating surface	surface de chauffe de la chaudière	42,5
d_i'	mm	Innendurchmesser (gerechnet)	inside diameter (calculated)	diamètre intérieur (calculé)	56,7
d_i	mm	Innendurchmesser (ausgeführt)	inside diameter (actual)	diamètre intérieur (réel)	57,5
d_n	mm	Nenndurchmesser	nominal diameter	diamètre nominal	57

D Niederdruck-Dampfheizung (Standrohre). — Low Pressure Steam Heating (Stand Pipes). — Chauffage à vapeur à basse pression (colonnes de sûreté).

F_k	m²	Kesselheizfläche	boiler heating surface	surface de chauffe de la chaudière	10,4
d_i'	mm	Innendurchmesser (gerechnet)	inside diameter (calculated)	diamètre intérieur (calculé)	68,0
d_i	mm	Innendurchmesser (ausgeführt)	inside diameter (actual)	diamètre intérieur (réel)	70,0
d_n	mm	Nenndurchmesser	nominal diameter	diamètre nominal	70

Ausführungsarten. — Methods of Construction. — Genres d'exécution.

WA	eine Leitung, die am Ausdehnungsgefäß unten mündet	one pipe line, to bottom of expansion tank	une conduite aboutissant au bas du vase d'expansion
WB	zwei Leitungen (Ausdehnung und Rücklauf)	two pipe lines (expansion and return)	deux conduites (expansion et retour)

Leitungsarten. — Nature of Pipe Lines. — Genres de conduites.

1	Sicherheits-Ausdehnungsleitungen	safety expansion lines	conduites d'expansion et de sûreté
2	Umgehungs- und Ausblaseleitungen	by-pass and blow-off lines	conduites de by-pass et de purge
3	Sicherheits-Rücklaufleitungen	safety return lines	tuyauteries de retour de sûreté

Preußische Ministerialvorschriften von 1925.

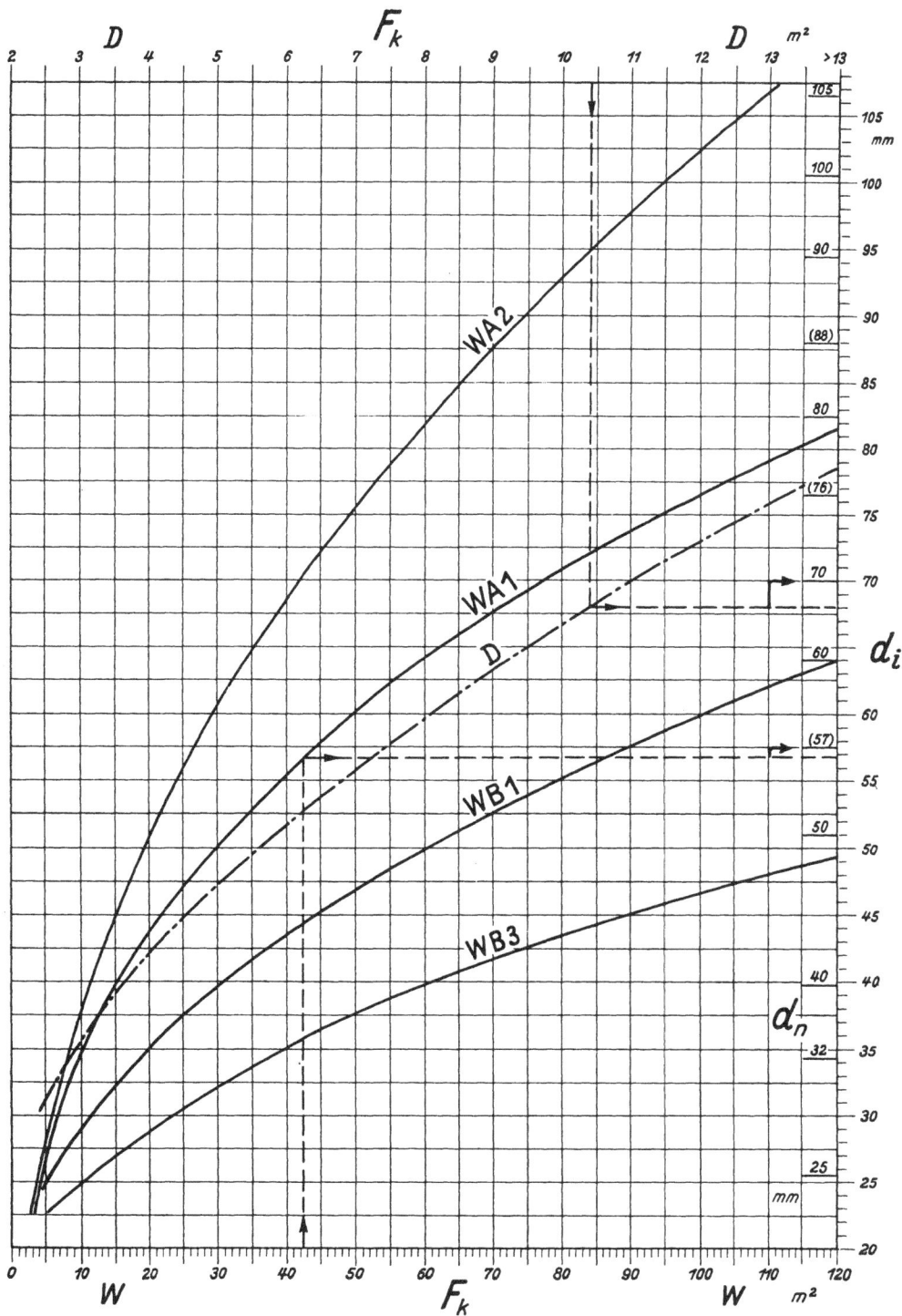

13

Schornstein-Zugstärke.
Chimney Draught.
Tirage de la cheminée.

h_{sch}	m	Schornsteinhöhe	height of chimney	hauteur de la cheminée	27,0
t_R	°C	Temperatur der Rauchgase	temperature of flue gases	température des fumées	120
P_{sch_0}	mm H_2O	Schornstein-Zugstärke (für 0° C Außentemperatur)	chimney draught (with outdoor temperature 0° C)	tirage de la cheminée (pour température extérieure de 0° C)	10,1
t_a	°C	Außentemperatur	outdoor temperature	température extérieure	$+10$
P_{sch_a}	mm H_2O	Zugstärkenänderung (für andere Außentemperaturen)	variation of draught (for other outdoor temperatures)	variation de tirage (pour les autres valeurs de la température extérieure)	1,25
P_{sch}	mm H_2O	Schornstein-Zugstärke	chimney draught	tirage de la cheminée	8,85

$$P_{sch} = h_{sch}\left[\gamma_{L_0}\frac{273}{t_a + 273} - \gamma_{R_0}\frac{273}{t_R + 273}\right]$$

$$P_{sch} = P_{sch_0} - P_{sch_a}$$

		für:	for:	pour:	
γ_{L_0}	$\dfrac{kg}{Nm^3}$	spezifisches Gewicht der Luft (für 0° C und 760 mm Hg)	density of air (at 0° C and 760 mm Hg)	poids spécifique de l'air (ramené à 0° C et 760 mm Hg)	1,293
γ_{R_0}	$\dfrac{kg}{Nm^3}$	spezifisches Gewicht der Rauchgase (für 0° C und 760 mm Hg)	density of flue gases (at 0° C and 760 mm Hg)	poids spécifique des fumées (ramené à 0° C et 760 mm Hg)	1,329

Gumz W., Feuerungstechnisches Rechnen. Leipzig 1931.

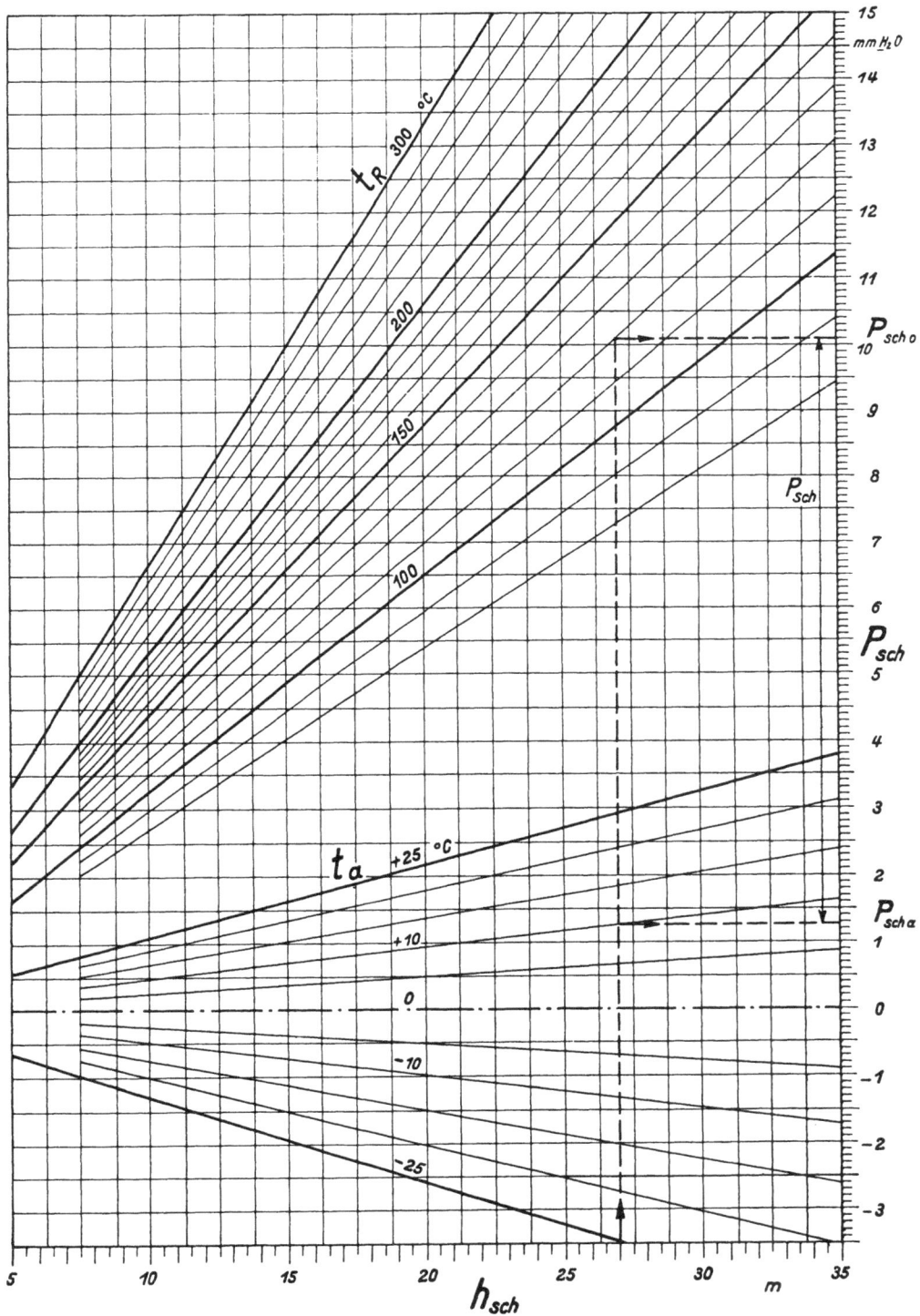

Schornstein-Querschnitt.
Chimney Area.
Section de la cheminée.

H_u	$\dfrac{\text{kcal}}{\text{kg}}$	unterer Heizwert (nur feste Brennstoffe)	net calorific value (solid fuels only)	pouvoir calorifique inférieur (combustibles solides seulement)	7000
n		Luftüberschußzahl	excess air ratio	coefficient d'excès d'air	2,0
V_{R_0}	$\dfrac{\text{Nm}^3}{\text{kg}}$	Rauminhalt der Rauchgase (für 0° C und 760 mm Hg)	specific volume of flue gases (at 0° C and 760 mm Hg)	volume spécifique des fumées (ramené à 0° C et 760 mm Hg)	15,5
t_R	°C	Temperatur der Rauchgase	temperature of flue gases	température des fumées	120
F_{sch}	cm²	Schornsteinquerschnitt	chimney area	section de la cheminée	600
M_B	$\dfrac{\text{kg}}{\text{h}}$	stündliche Brennstoffmenge	quantity of fuel per hour	quantité de combustible par heure	20,0
w_R	$\dfrac{\text{m}}{\text{s}}$	Rauchgasgeschwindigkeit	velocity of flue gases	vitesse des fumées	2,05

$$w_R = \frac{10000}{3600} \cdot \frac{M_B}{F_{sch}} \cdot \frac{273 + t_R}{273} \cdot V_{R_0}$$

Rosin-Fehling, It-Diagramm der Verbrennung. Berlin 1929.

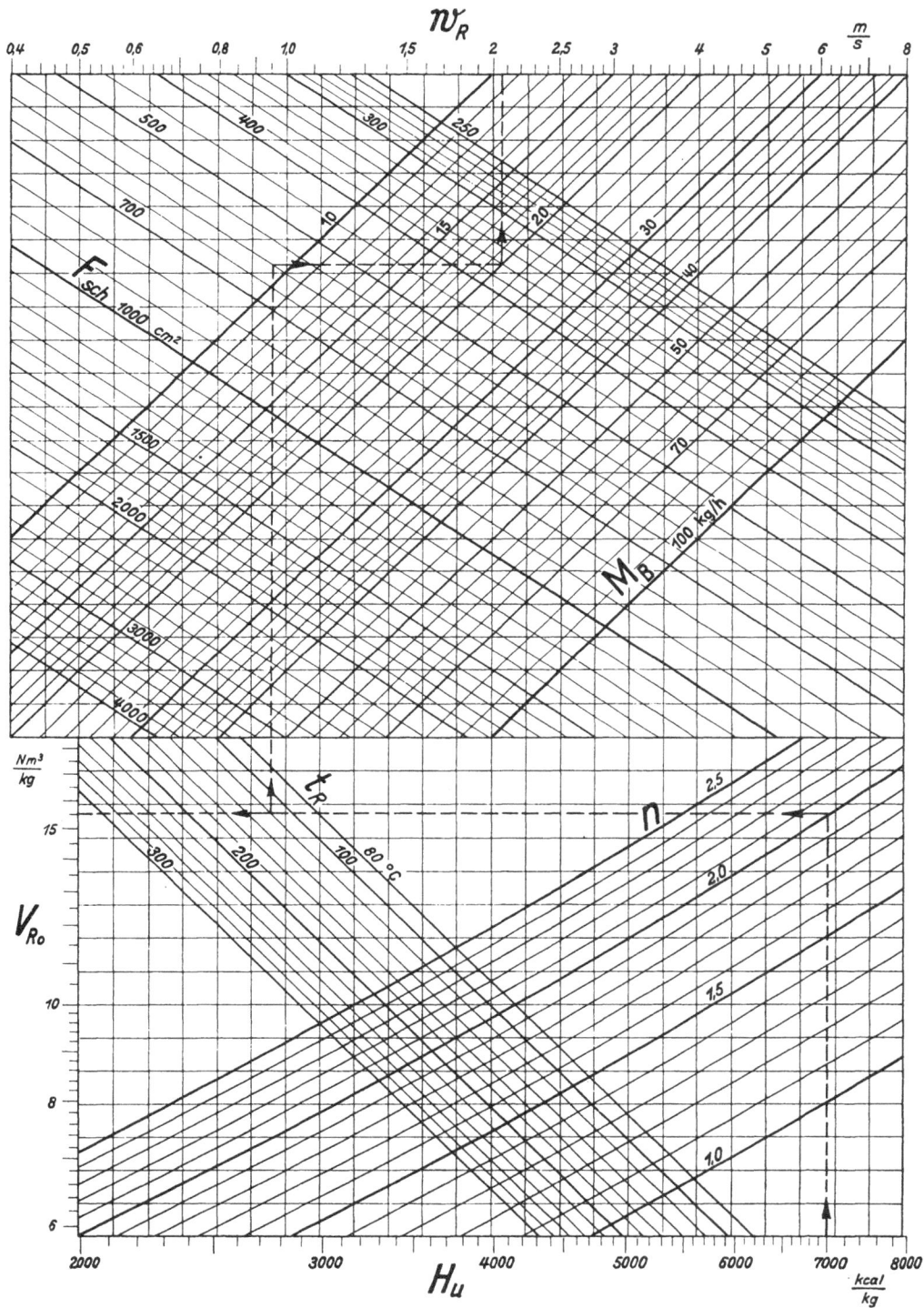

15

Zugverluste im Schornstein.
Loss of Draught in Chimney.
Pertes de tirage dans la cheminée.

I. Geschwindigkeitsverlust. — Loss due to Velocity. — Perte de charge cinétique.

①

w_R	$\dfrac{m}{s}$	Rauchgasgeschwin-digkeit	velocity of flue gases	vitesse des fumées	2,05
t_R	°C	Temperatur der Rauchgase	temperature of flue gases	température des fu-mées	120
Z_w	mm H$_2$O	Geschwindigkeits-verlust	loss due to velocity	perte de charge ci-nétique	0,2

$$Z_w = \frac{w^2}{2\,g} \cdot \gamma_{R_0} \cdot \frac{273}{t_R + 273}$$

II. Reibungsverlust. — Loss due to Friction. — Perte de charge par frottement.

②

w_R	$\dfrac{m}{s}$	Rauchgasgeschwin-digkeit	velocity of flue gases	vitesse des fumées	2,05
F_{sch}	cm²	Schornsteinquer-schnitt	chimney area	section de la chemi-née	600
t_R	°C	Temperatur der Rauchgase	temperature of flue gases	température des fu-mées	120
z_r	$\dfrac{mm\ H_2O}{m}$	Reibungsverlust (je 1 m Schornstein-höhe)	loss due to friction (per metre of chimney height)	perte par frotte-ment (par m de hauteur de che-minée)	0,038
h_{sch}	m	Schornsteinhöhe	chimney height	hauteur de la che-minée	27,0
Z_r	mm H$_2$O	Reibungsverlust (im Schornstein)	loss due to friction	perte par frotte-ment	1,03
Z_f	mm H$_2$O	Mittelwert des Rei-bungsverlustes (im Fuchs)	mean value of loss due to friction (in flue)	valeur moyenne de la perte par frot-tement (dans le carnean)	1,0
Z_{ges}	mm H$_2$O	gesamter Zugver-lust	total loss of draught	perte de charge to-tale	2,23
P_{res}	mm H$_2$O	wirksame Zugstärke	effective chimney draught	tirage efficace de la cheminée	6,57

$$Z_r = h_{sch} \cdot z_r$$
$$Z_{ges} = Z_w + Z_r + Z_f$$
$$P_{res} = P_{sch} - Z_{ges}$$

Noelpp, Schornstein-Berechnung und Schornstein-Ausführung. Gesundh.-Ing. Bd. 57 (1934), S. 587.

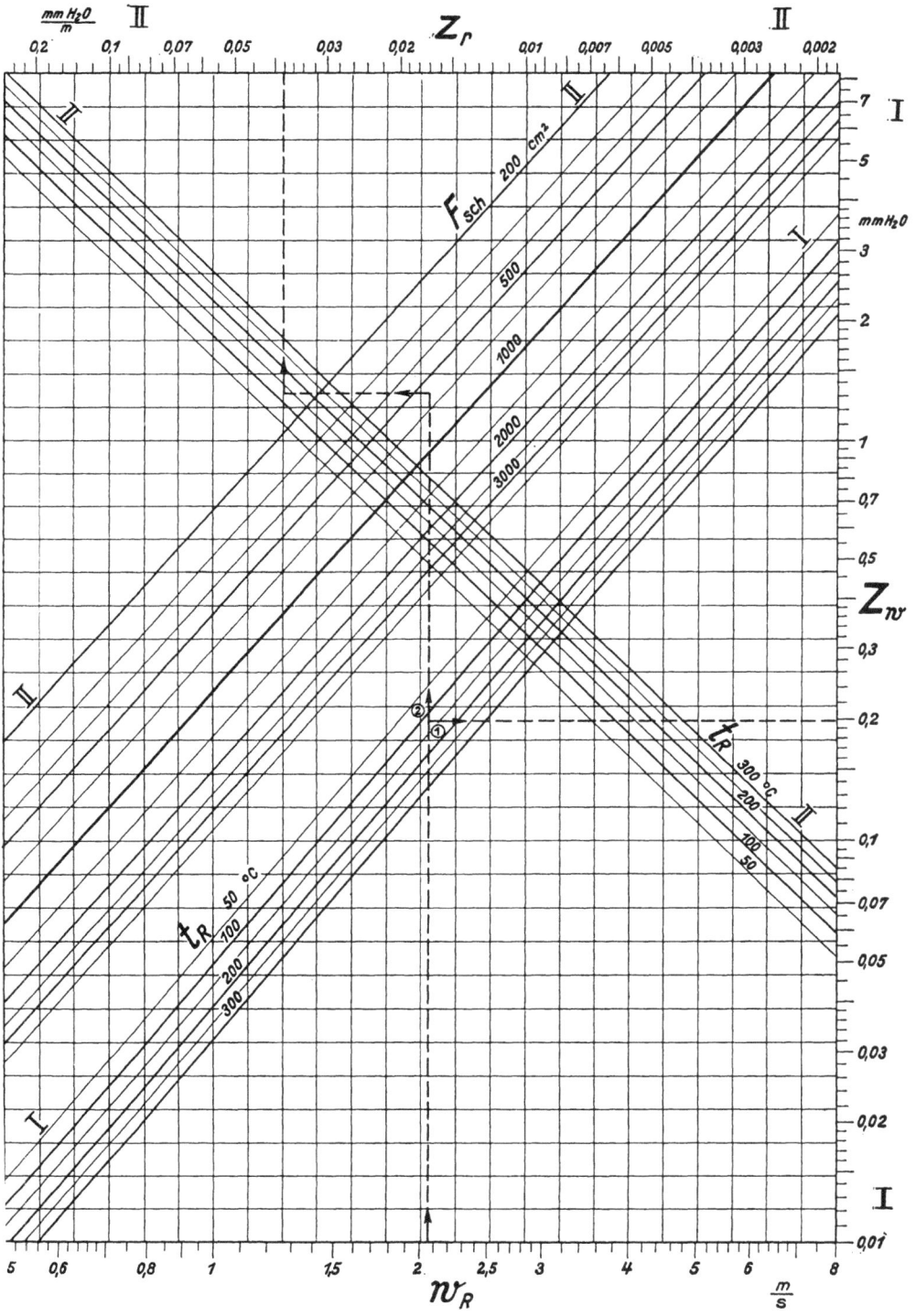

$\frac{mm\,H_2O}{m}$ II Z_r II

0,2   0,1   0,07   0,05   0,03   0,02   0,01   0,007   0,005   0,003   0,002

II

II

II

I

7
5

$mm\,H_2O$

3

2

F_{sch} 200 cm²

500

1000

2000

3000

1

0,7

0,5

Z_{nv}

0,3

II

0,2

t_R 300 °C

200

100

II

50

0,1

t_R 50 °C

100

200

300

0,07

0,05

0,03

0,02

I

I

0,01

5   0,6   0,8   1   1,5   2   2,5   3   4   5   6   8

w_R $\frac{m}{s}$

16

Rohrabmessungen.
Pipe Dimensions.
Dimensions des tuyaux.

d_n	mm	Nenndurchmesser	nominal diameter	diamètre nominal	100
d_i	mm	Innendurchmesser	inside diameter	diamètre intérieur	100,5
δ	mm	Wandstärke	wall thickness	epaisseur de paroi	3,75
d_a	mm	Außendurchmesser	outside diameter	diamètre extérieur	108,0
F_i	cm²	lichter Querschnitt	inside cross-sectional area	section intérieure nette	79,0
V_i	$\dfrac{1}{m}$	Rauminhalt (je 1 m Rohrlänge)	capacity (per metre-run of pipe)	contenance (par m de tuyau)	7,9
F_E	cm²	Eisenquerschnitt	cross-sectional area of iron	section de fer (du tuyau)	12,3
G_E	$\dfrac{kg}{m}$	Eisengewicht (je 1 m Rohrlänge)	weight of iron (per metre-run of pipe)	poids de fer (par m de tuyau)	9,8

$$d_a = d_i + 2\,\delta$$

$$F_i = \left(\frac{d_i}{10}\right)^2 \cdot \frac{\pi}{4}$$

$$V_i = \left(\frac{d_i}{10}\right)^2 \cdot \frac{\pi}{40}$$

DIN 4701. — Recknagel. — Rietschel.

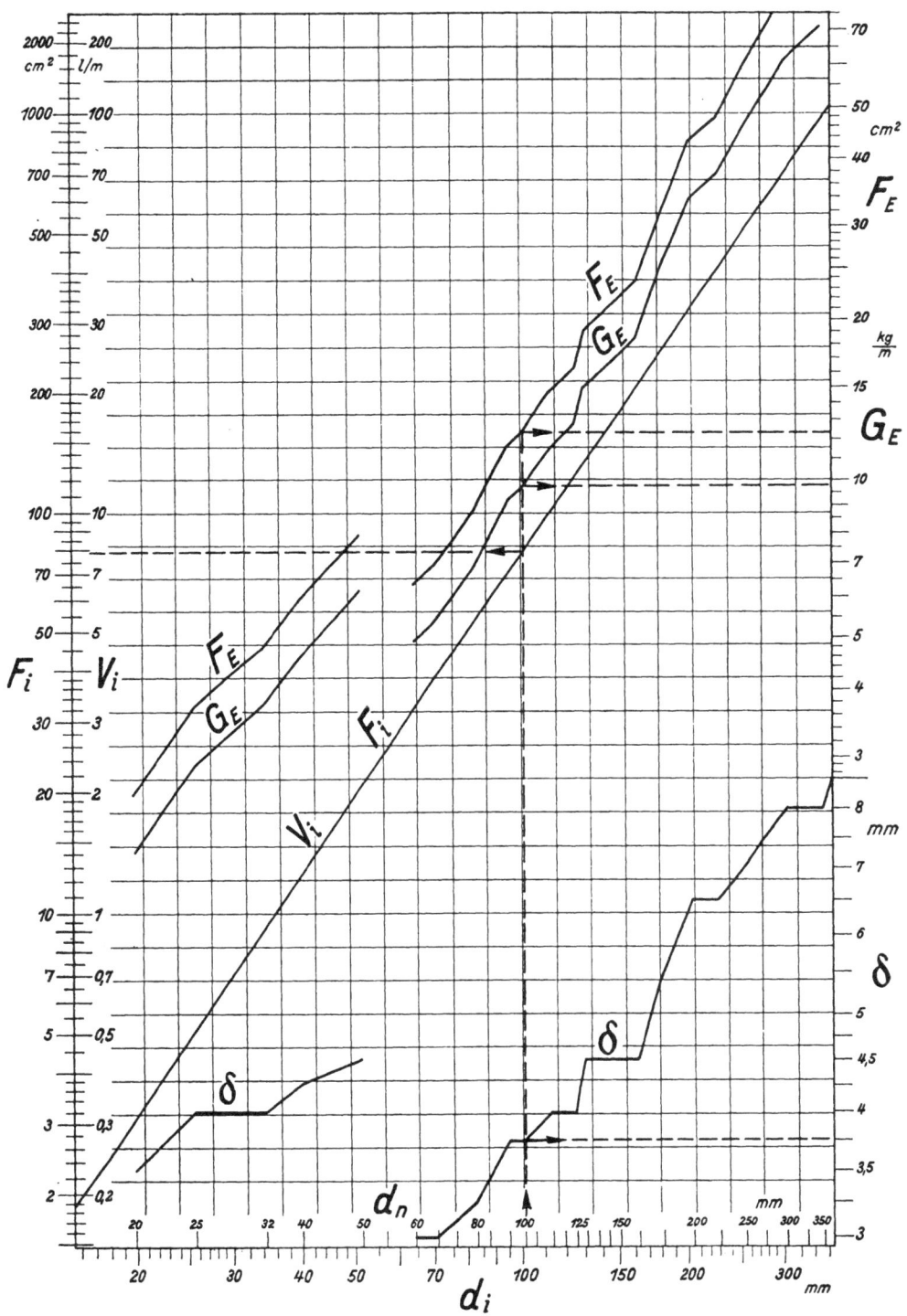

Rohroberfläche (mit und ohne Wärmeschutz).
Pipe Surface (with and without Lagging).
Surface des tuyaux (avec et sans calorifuge).

d_n	mm	Nenndurchmesser	nominal diameter	diamètre nominal	100
d_i	mm	Innendurchmesser	inside diameter	diamètre intérieur	100,5
δ_J	mm	Stärke des Wärmeschutzes	thickness of insulation	épaisseur du revêtement calorifuge	40,0
F_J	$\dfrac{m^2}{m}$	Rohroberfläche (je 1 m Rohrlänge)	pipe surface (per metre run of pipe)	surface du tuyau (par m de longueur)	0,59

Recknagel.

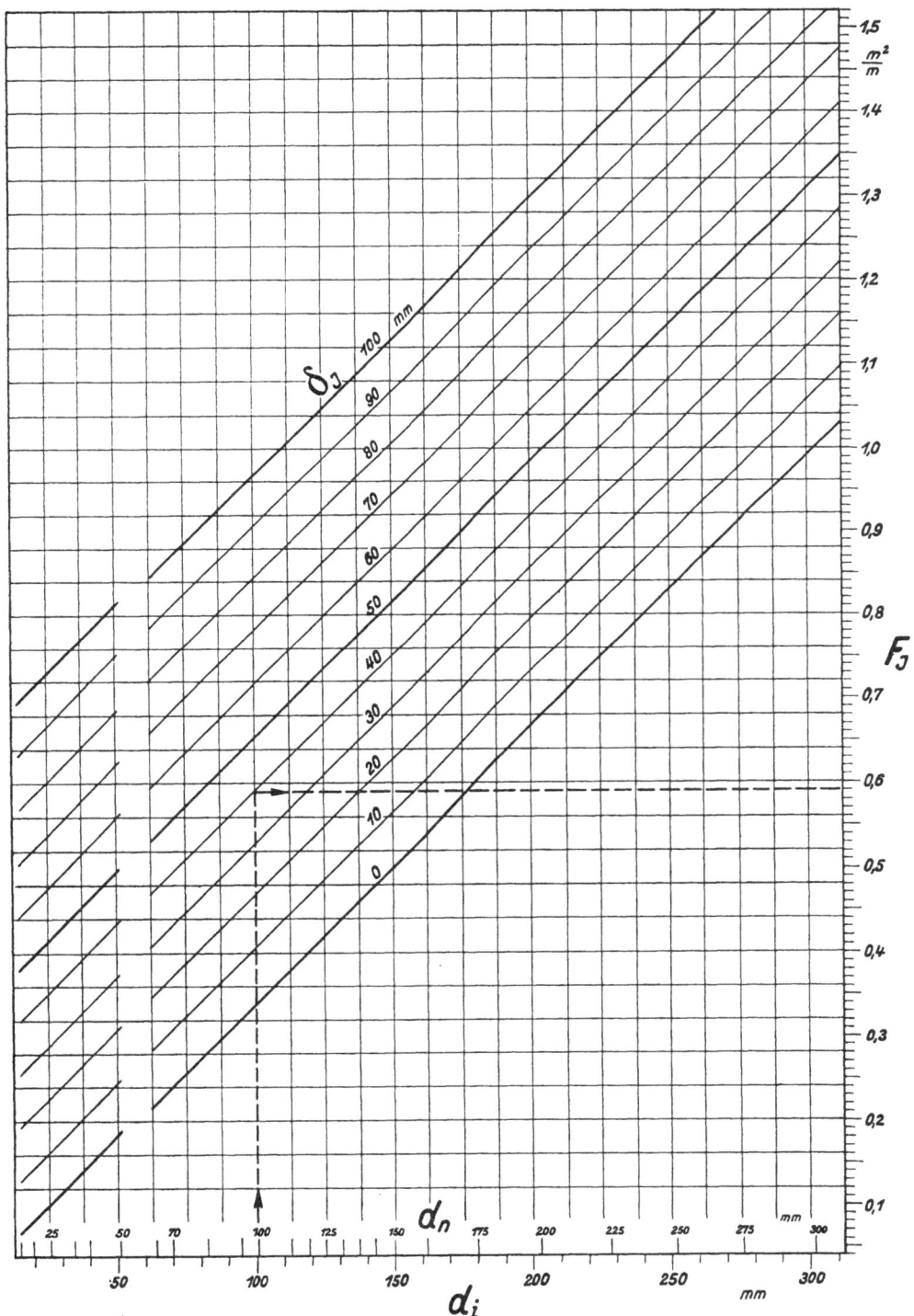

Wärmeinhalt und Strömungsgeschwindigkeit.
Heat Content and Velocity of Flow.
Quantité de chaleur et vitesse de circulation.

I. Warmwasser. — Hot Water. — Eau chaude.

d_n	mm	Nenndurchmesser	nominal diameter	diamètre nominal	80
d_i	mm	Innendurchmesser	inside diameter	diamètre intérieur	82,5
w_W	$\dfrac{m}{s}$	Strömungs- geschwindigkeit des Wassers	velocity of flow of water	vitesse de circu- lation	0,10
t_W	°C	Wassertemperatur	temperature of water	température de l'eau	90,0
J_W	$\dfrac{1000\ kcal}{h}$	stündlicher Wärme- inhalt des strö- menden Wassers	heat content of wa- ter flow per hour	quantité de chaleur horaire dans l'eau de circulation	167

$$J_W = 3600 \cdot w_W \cdot \frac{\pi \cdot d_i^2}{4} \cdot i_W \cdot \gamma_W$$

II. Niederdruckdampf. — Low Pressure Steam. — Vapeur à basse pression.

d_n	mm	Nenndurchmesser	nominal diameter	diamètre nominal	50
d_i	mm	Innendurchmesser	inside diameter	diamètre intérieur	51,0
w_D	$\dfrac{m}{s}$	Strömungs- geschwindigkeit des Dampfes	velocity of flow of steam	vitesse d'écoule- ment de la vapeur	15
p_D	ata	Dampfdruck	steam pressure	pression de la va- peur	1,4
J_D	$\dfrac{1000\ kcal}{h}$	stündlicher Wärme- inhalt des strö- menden Dampfes	heat content of steam flow per hour	quantité de chaleur horaire emportée par la vapeur	56

$$J_D = 3600 \cdot w_D \cdot \frac{\pi \cdot d_i^2}{4} \cdot i_D \cdot \gamma_D$$

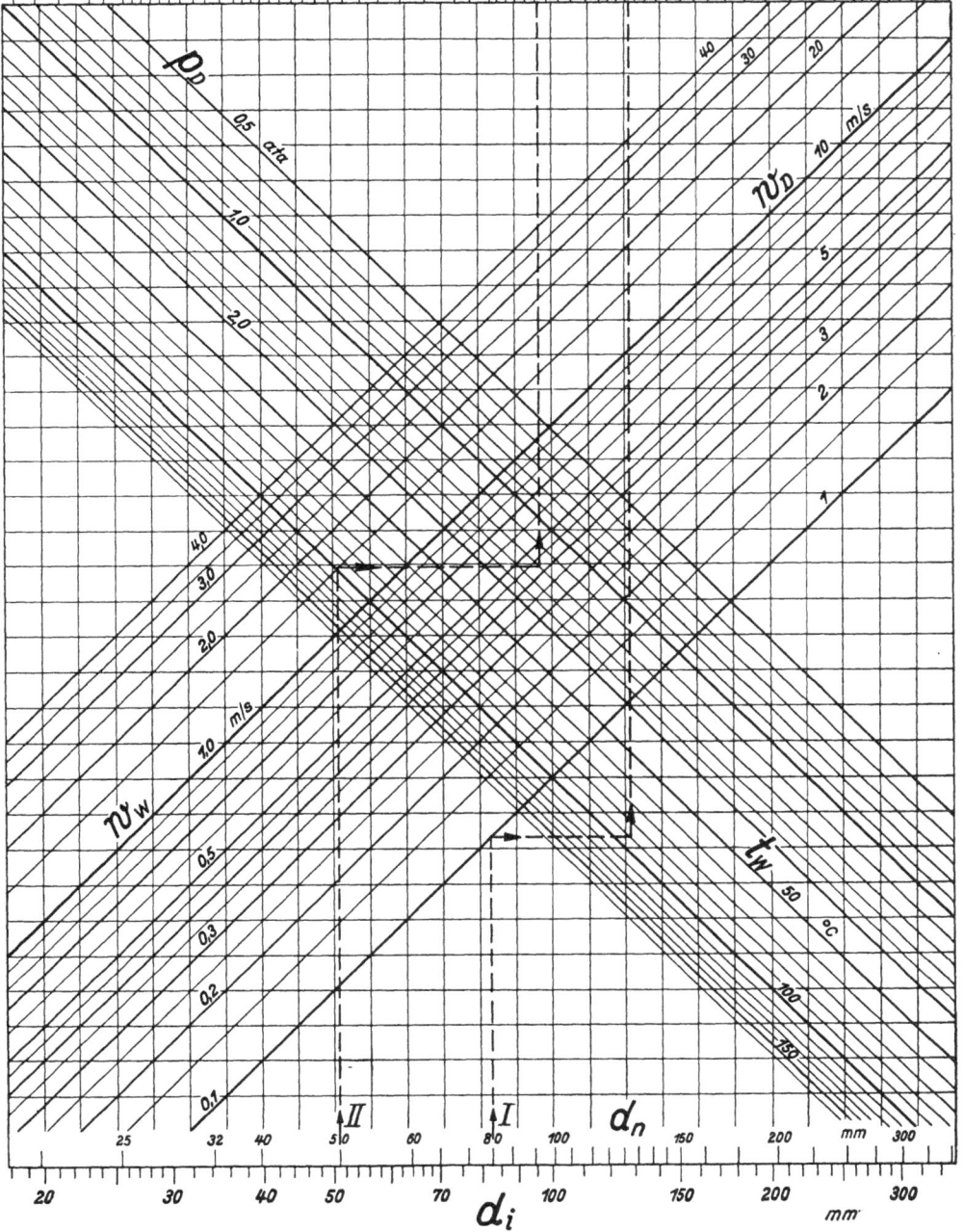

4*

19

Wirksamer Druckunterschied (in Schwerkraftheizungen).
Effective Pressure Difference (in Gravity-Hot Water Systems).
Différence de pression efficace (dans les chauffages à eau chaude par gravité).

						①	②
t_{W_u}	°C	Wassertemperatur im Fallstrang	temperature of water in fall pipe	température de l'eau dans la colonne de retour		70	75
t_{W_o}	°C	Wassertemperatur im Steigstrang	temperature of water in rising pipe	température de l'eau dans la colonne montante		90	95
$\varDelta P$	$\dfrac{\text{mm } H_2O}{\text{m}}$	wirksamer Druckunterschied	effective difference of pressure	différence de pression efficace		12,5	13,0

Rietschel.

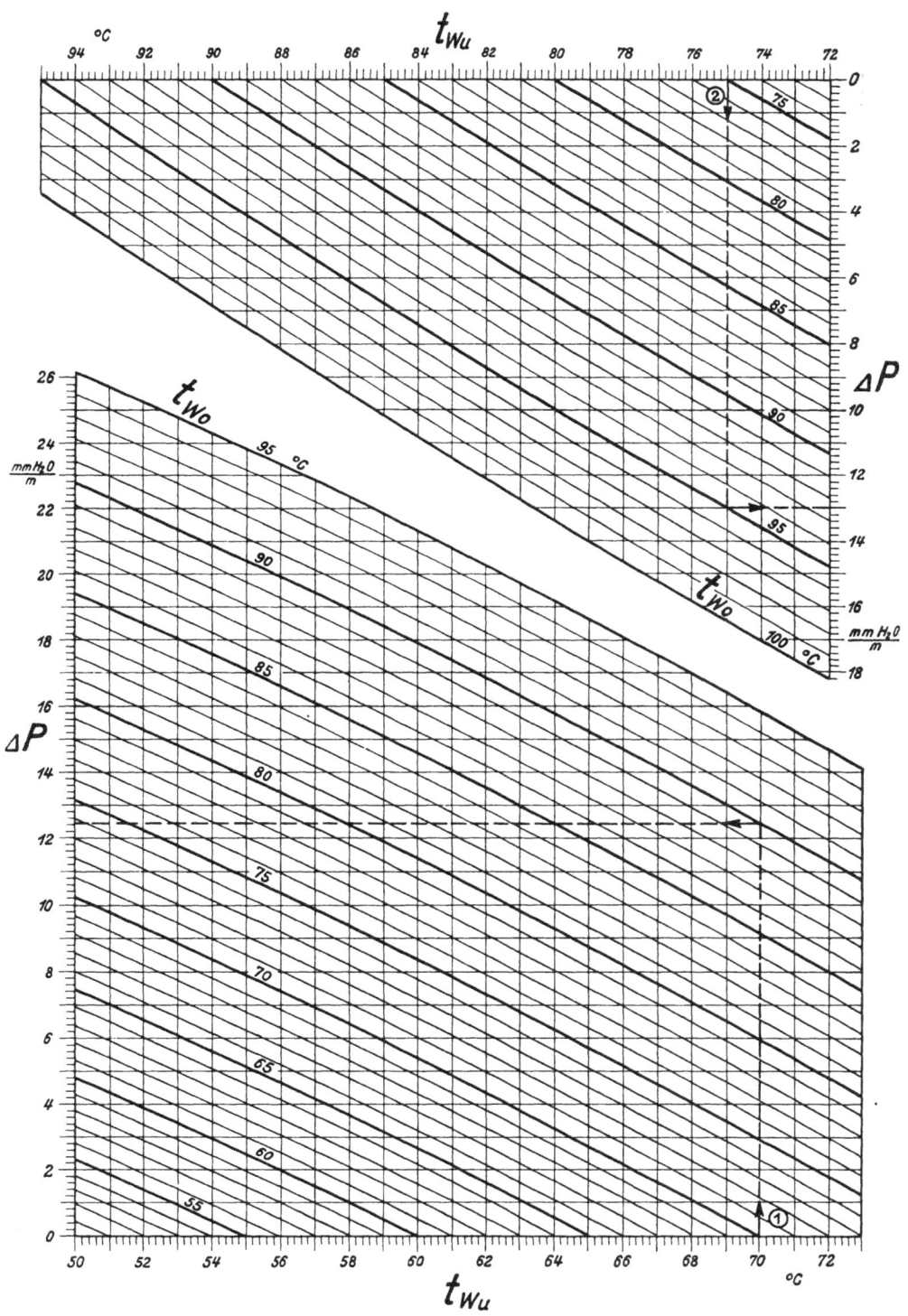

20

Wärmeverlust isolierter Rohrleitungen.
Loss of Heat from Insulated Pipe Lines.
Pertes thermiques des tuyauteries calorifugées.

						①	②
λ	$\dfrac{kcal}{m\,h\,^oC}$	Wärmeleitzahl	thermal conductivity	coefficient de conductibilité thermique		0,12	0,055
d_n	mm	Nenndurchmesser	nominal diameter	diamètre nominal		100	40
δ_J	mm	Stärke der Isolierung	thickness of insulation	épaisseur du revêtement calorifuge		50	40
q_J	$\dfrac{kcal}{m\,h\,^oC}$	Wärmeverlust (je 1 m Rohrlänge)	heat loss (per metre-run of pipe)	déperdition de chaleur (par m de tuyau)		0,895	0,30

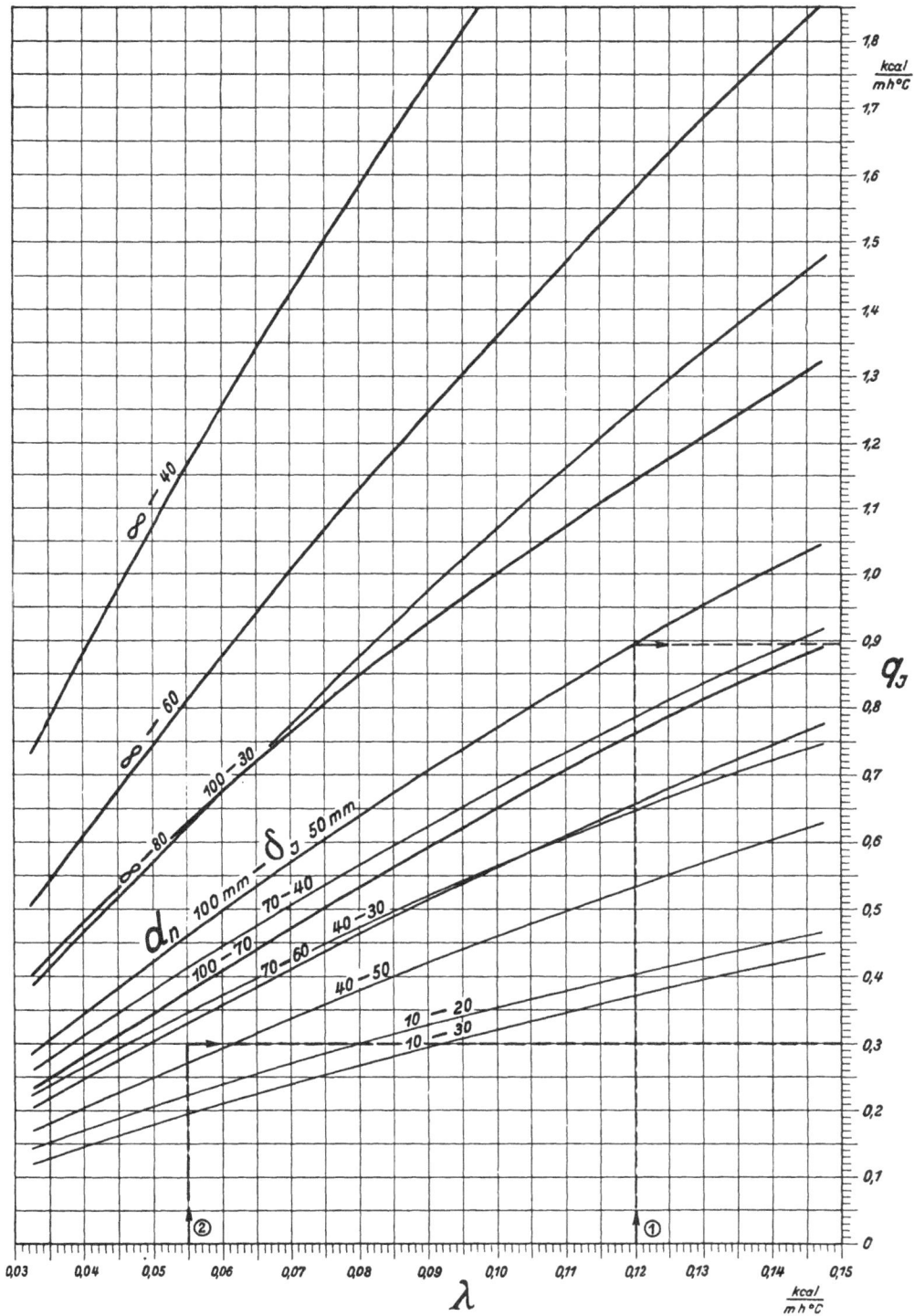

21

Abkühlung durch Wärmeverluste.
Temperature Drop by Heat Losses.
Refroidissement par pertes de chaleur.

q_J	$\dfrac{\text{kcal}}{\text{m h}\,^\circ\text{C}}$	Wärmeverlust (je 1 m Rohrlänge)	heat loss (per metre run of pipe)	déperdition de chaleur (par m de conduite)	0,3
l	m	Länge der Rohrleitung	length of pipe line	longueur de la tuyauterie	4,5
M_W	$\dfrac{l}{h}$	stündliche Wassermenge	quantity of water per hour	quantité d'eau par heure	350
t_{W_E}	$^\circ$C	Eintrittstemperatur des Wassers	inlet temperature of water	température d'entrée de l'eau	88
t_L	$^\circ$C	Temperatur der umgebenden Luft	room temperature	température du local	35
$t_{W_E}-t_{W_A}$	$^\circ$C	Abkühlung des Wassers (beim Durchfließen der Rohrleitung)	temperature drop of water in pipes	refroidissement de l'eau au passage dans la conduite	0,2
t_{W_A}	$^\circ$C	Austrittstemperatur des Wassers	outlet temperature of water	température de sortie de l'eau	87,8

$$t_{W_E} - t_{W_A} = \frac{q_J \cdot l \cdot (t_{W_E} - t_L)}{M_W}$$

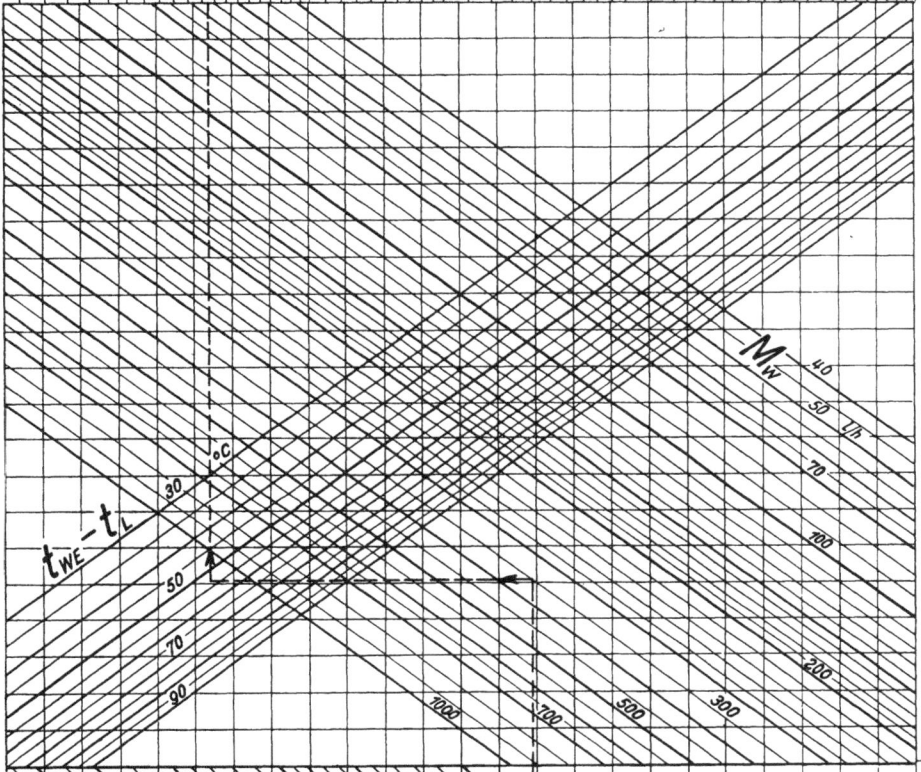

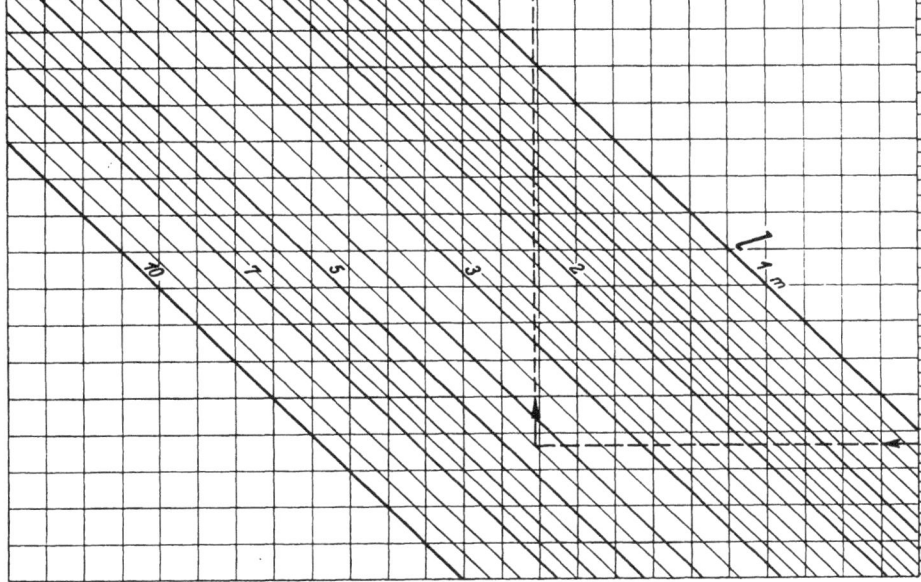

22

Wärmeleistung von Schwerkraft-Warmwasserheizungen (für 20⁰ C Temperaturgefälle).
Heat Output of Gravity-Hot Water Systems (for 20⁰ C Temperature Drop).
Pouvoir de chauffe des installations de chauffage à eau chaude par gravité (pour une chute de température de 20⁰ C).

①

p_l	$\dfrac{\text{mm H}_2\text{O}}{\text{m}}$	verfügbares Druckgefälle (je 1 m Rohrlänge)	available pressure drop (per metre run of pipe)	chute de pression disponible (par m de conduite)	0,6
Q_h	$\dfrac{\text{kcal}}{\text{h}}$	notwendige Wärmeleistung	requisite heat output	quantité de chaleur nécessaire	18000
d_i	mm	Innendurchmesser (berechnet)	internal diameter (calculated)	diamètre intérieur (calculé)	48,5

②

d_n	mm	Nenndurchmesser	nominal diameter	diamètre nominal	50
d_i'	mm	Innendurchmesser (ausgeführt)	internal diameter (actual)	diamètre intérieur (réel)	51,0
p_l	$\dfrac{\text{mm H}_2\text{O}}{\text{m}}$	verbrauchtes Druckgefälle (je 1 m Rohrlänge)	pressure drop utilised (per metre run of pipe)	chute de pression utilisée (par m de conduite)	0,47

Rietschel.

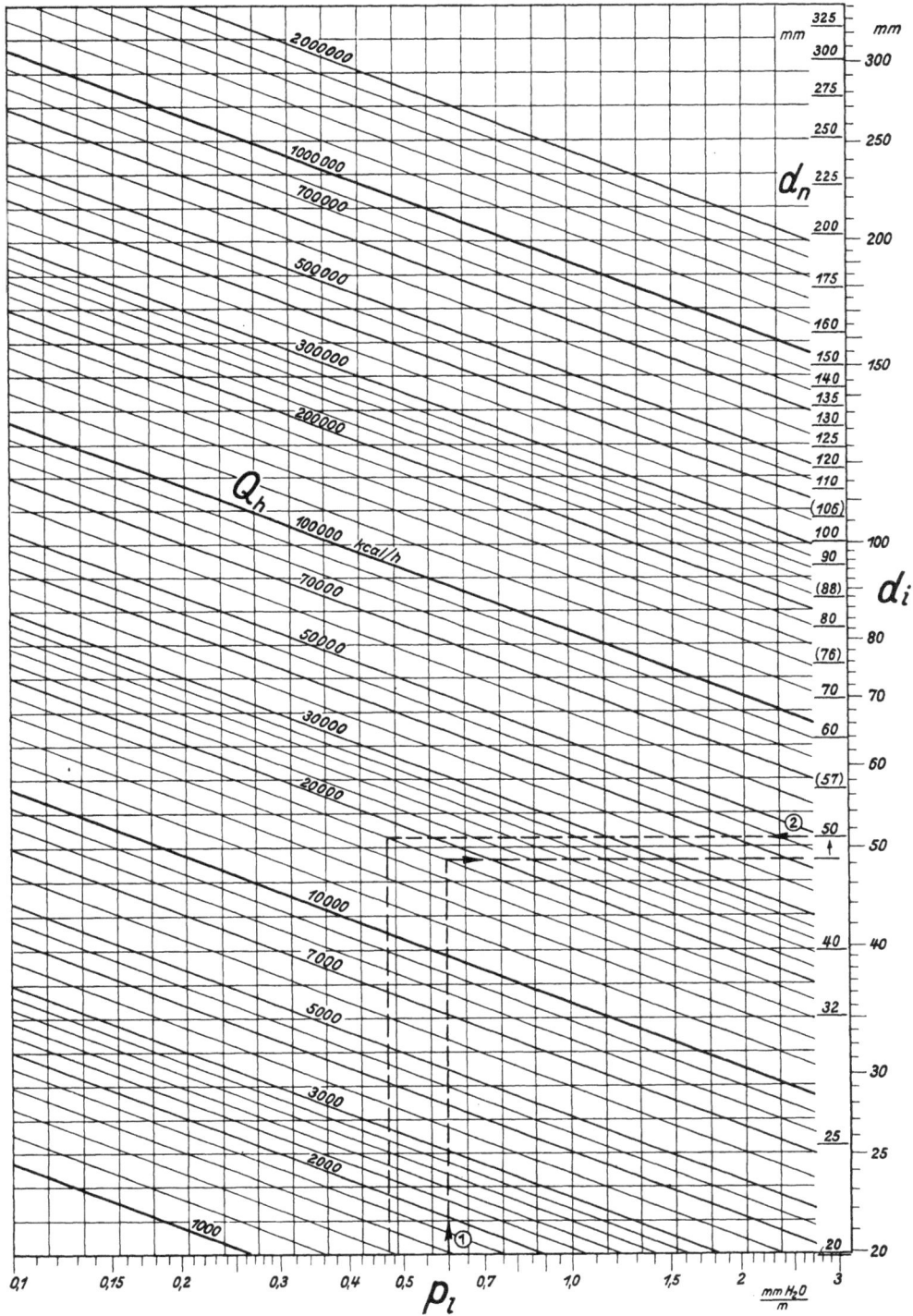

23

Umrechnung der Wärmeleistung (für beliebiges Temperaturgefälle).
Conversion of Heat Output (for Given Temperature Drop).
Détermination du pouvoir de chauffe (pour des différences de température quelconques).

$t_v - t_r$	°C	Temperaturgefälle (zwischen Vor- und Rücklauf)	temperature difference between water in flow and in return	différence de température entre les canalisations d'amenée et de retour	12,0
Q_t	$\dfrac{\text{kcal}}{\text{h}}$	Wärmeleistung (bei beliebigem Temperaturgefälle)	heat output (for given temperature drop)	quantité de chaleur émise (pour une chute de température quelconque)	5500
Q_h	$\dfrac{\text{kcal}}{\text{h}}$	Wärmeleistung (bezogen auf 20° Temperaturgefälle)	heat output (referred to 20° C temperature drop)	quantité de chaleur émise (pour une chute de température de 20° C)	9200

$$Q_h = Q_t \cdot \frac{20}{t_v - t_r}$$

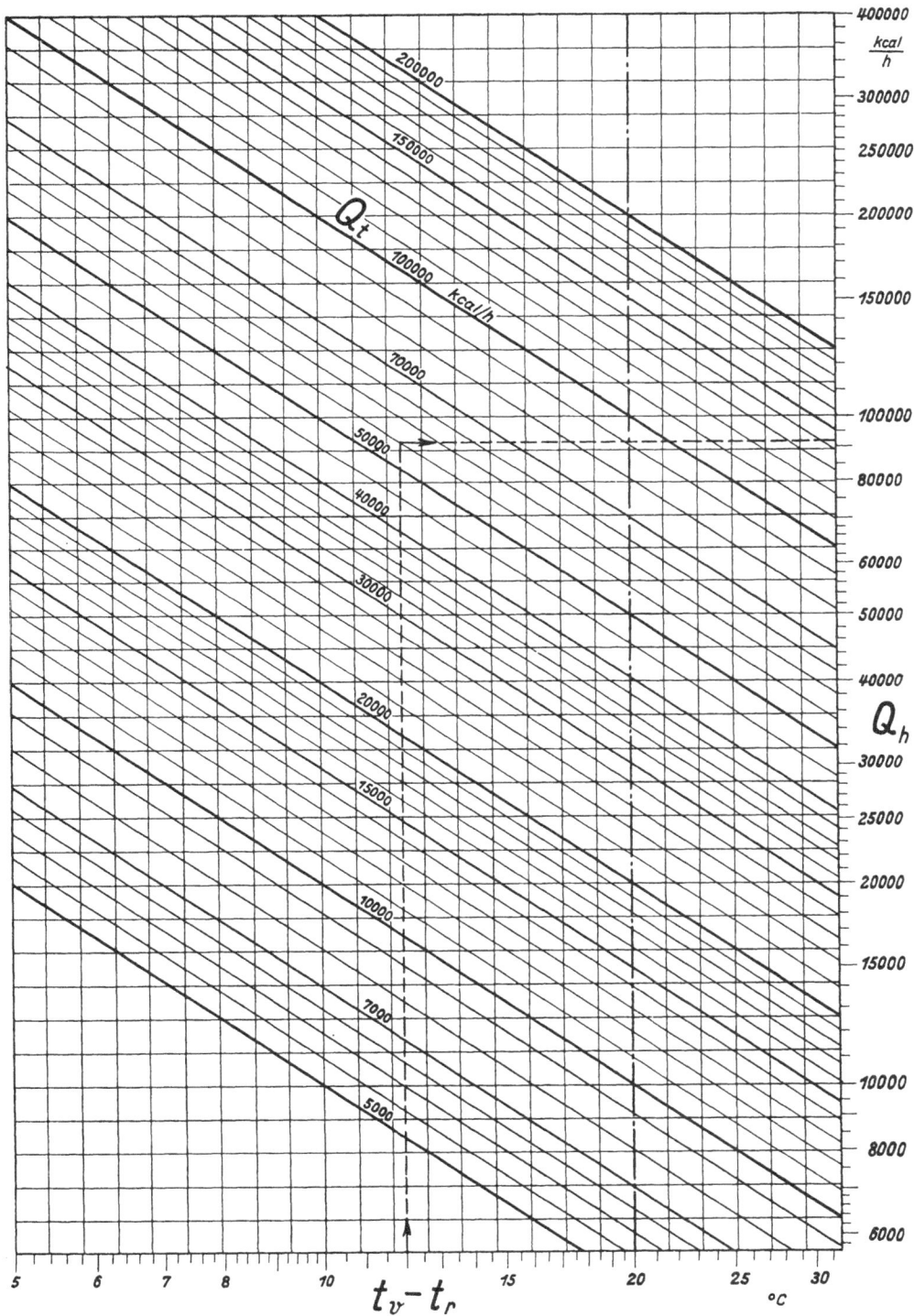

Strömungsgeschwindigkeit in Schwerkraft-Warmwasserheizungen.
Velocity of Flow in Gravity-Hot Water Systems.
Vitesse de circulation dans les installations de chauffage à eau chaude par gravité.

d_n	mm	Nenndurchmesser	nominal diameter	diamètre nominal	50
d_i	mm	Innendurchmesser	internal diameter	diamètre intérieur	51,0
p_l	$\dfrac{\text{mm } H_2O}{\text{m}}$	verfügbares Druck-gefälle (je 1 m Rohrlänge)	available pressure drop (per metre run of pipe)	chute de pression disponible (par m de conduite)	0,47
w_w	$\dfrac{\text{m}}{\text{s}}$	Strömungs-geschwindigkeit des Wassers	velocity of flow of water	vitesse de circu-lation de l'eau	0,126

Rietschel.

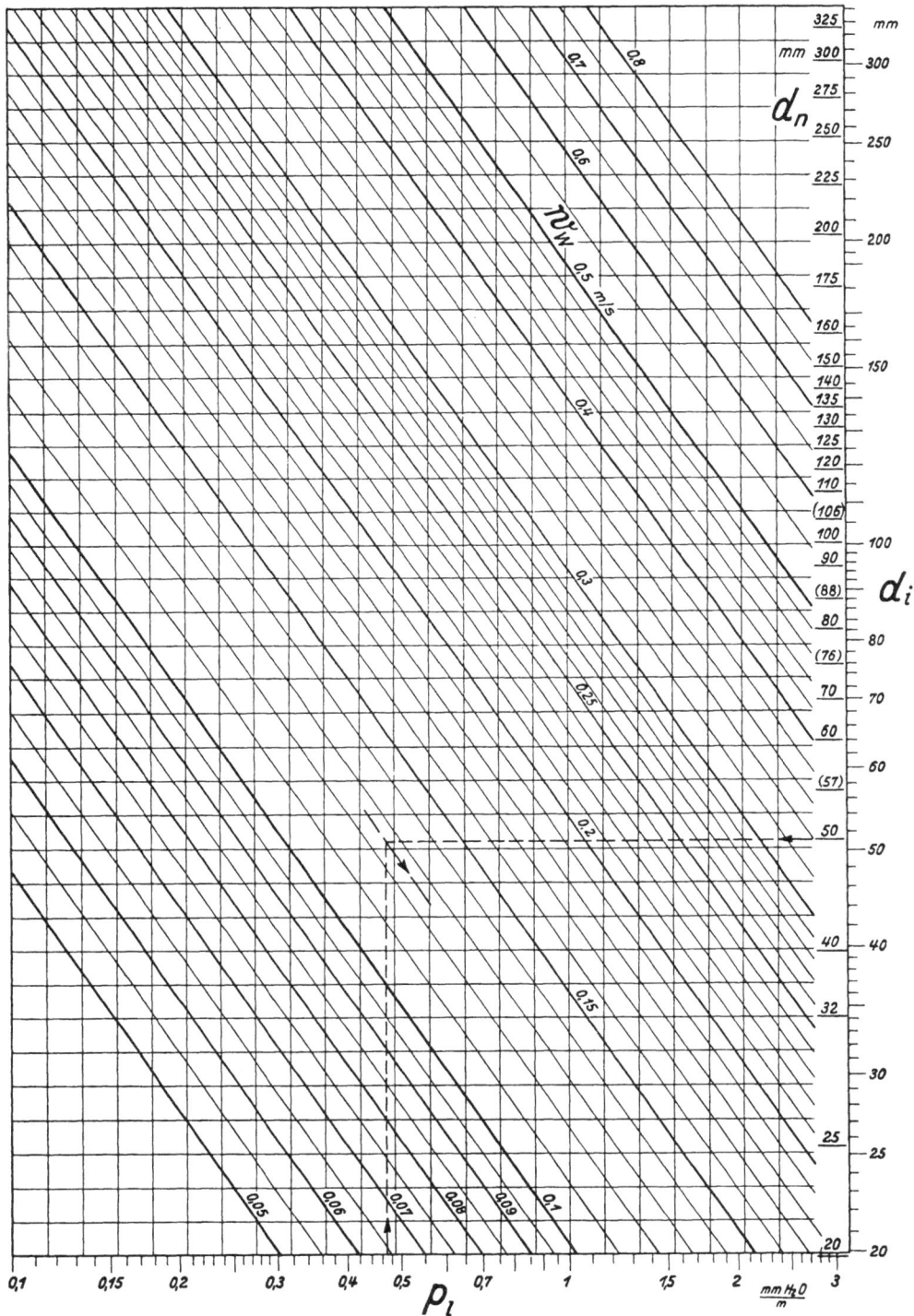

25

Einzelwiderstände der Rohrleitung.
Individual Resistances of Pipe Line.
Résistances locales de la tuyauterie.

		Heizmittel	heating Medium	agent de chauffage	① W I	② W II	③ D
$w_W w_D$	$\dfrac{\mathrm{m}}{\mathrm{s}}$	Strömungs-geschwindigkeit des Wassers bzw. Dampfes	velocity of flow of water or steam	vitesse de cir-culation de l'eau ou de la vapeur	0,12	0,89	18,0
$\Sigma\zeta$		Gesamtbeiwert der Einzelwider-stände	overall coefficient of individual re-sistances	coefficient glo-bal des ré-sistances lo-cales	4,5	3,0	2,0
$Z_W Z_D$	mm H_2O	Druckabfall in den Einzelwiderstän-den	pressure drop due to individual re-sistances	chute de pres-sion due aux résistances locales	3,2	118	21,0

Heizmittelarten. — Heating Medium. — Agent de chauffage.

W	Warmwasser	hot water	eau chaude
D	Niederdruckdampf	low pressure steam	vapeur à basse pression

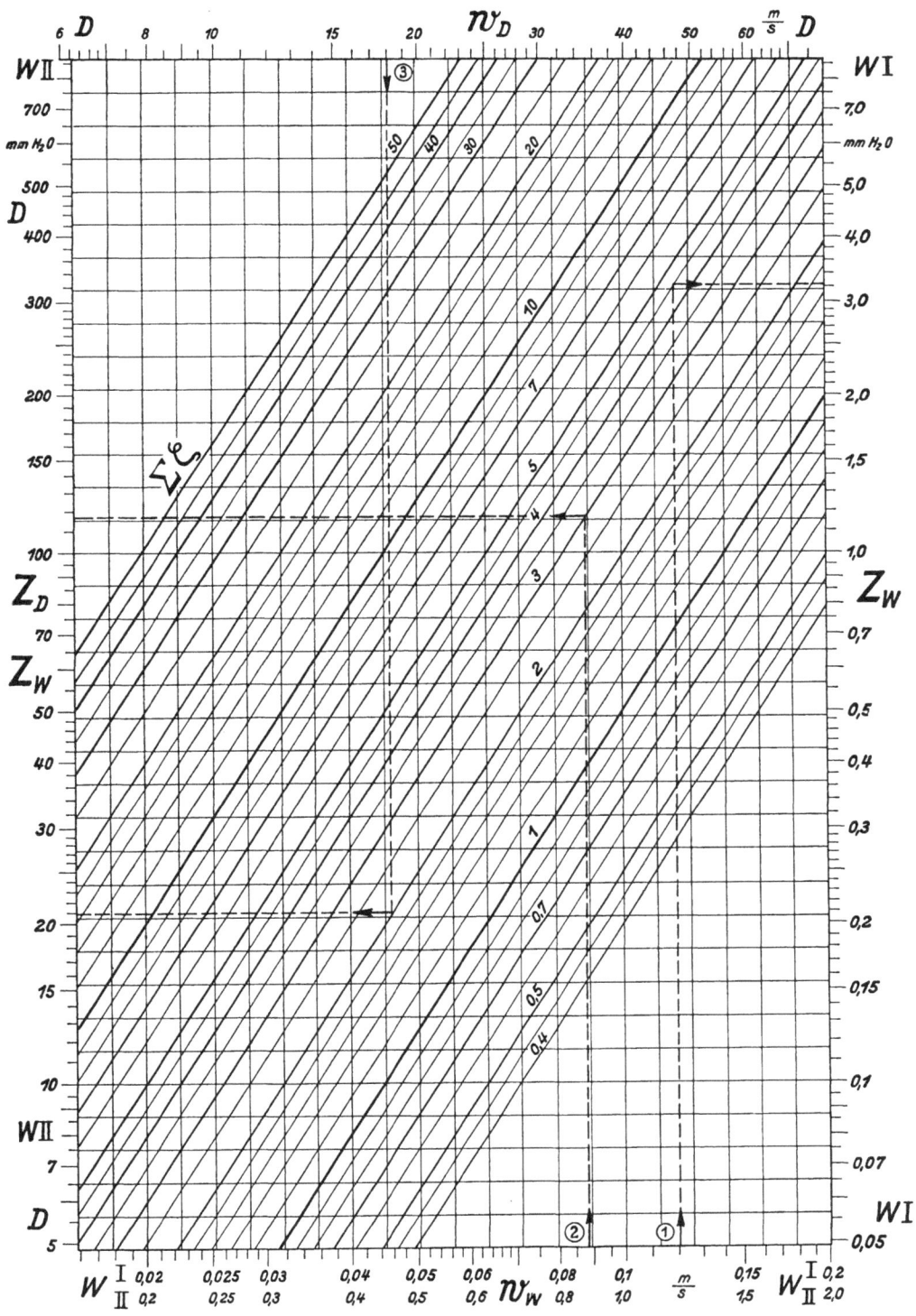

26

Wärmeleistung von Pumpen-Warmwasserheizungen (für 20° C Temperaturgefälle).
Heat Output of Forced Circulation Hot Water Systems (for 20° C Temperature Drop).
Pouvoir de chauffe des installations de chauffage à eau chaude avec circulation par pompe (pour différence de température de 20° C).

d_n	mm	Nenndurchmesser	nominal diameter	diamètre nominal	70
d_i	mm	Innendurchmesser	internal diameter	diamètre intérieur	70,0
Q_h	$\dfrac{\text{kcal}}{\text{h}}$	notwendige Wärme-leistung	requisite heat output	quantité de chaleur nécessaire par heure	240 000
p_l	$\dfrac{\text{mm } H_2O}{\text{m}}$	verfügbares Druck-gefälle (je 1 m Rohrlänge)	available pressure drop (per metre run of pipe)	chute de pression disponible (par m de conduite)	10,7

Rietschel.

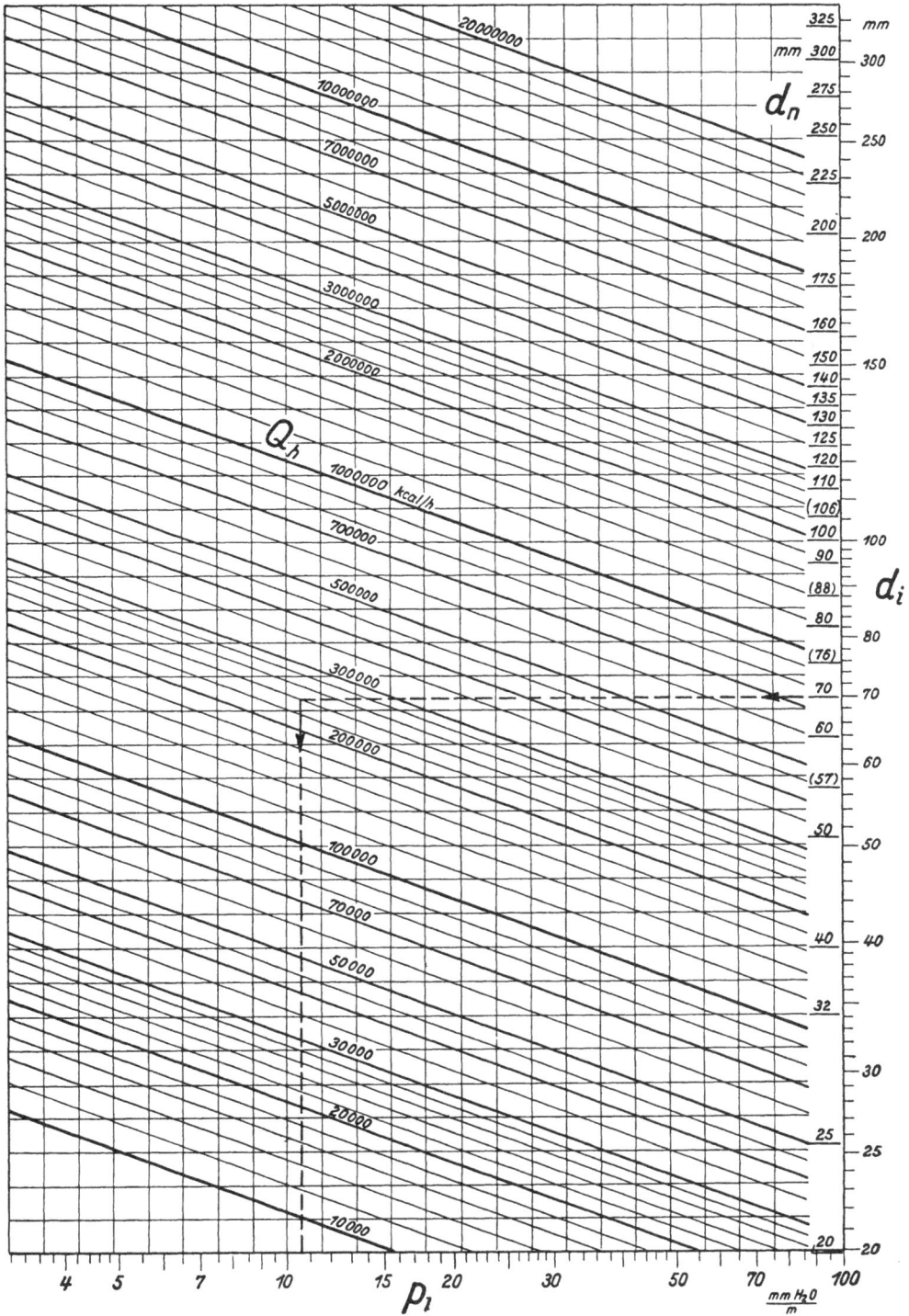

20000000

10000000

7000000

5000000

3000000

2000000

Q_h

1000000 kcal/h

700000

500000

300000

200000

100000

70000

50000

30000

20000

10000

d_n

mm

mm

325
300
275
250
225
200
175
160
150
140
135
130
125
120
110
(106)
100
90
(88)
80
(76)
70
60
(57)
50
40
32
25
20

d_i

mm
300
250
200
150
100
80
70
60
50
40
30
25
20

p_l $\frac{mm\,H_2O}{m}$

4   5   7   10   15   20   30   50   70   100

5*

27

Strömungsgeschwindigkeit in Pumpen-Warmwasserheizungen.
Velocity of Flow in Forced Circulation-Hot Water Systems.
Vitesse de circulation dans les installations de chauffage à eau chaude avec circulation par pompe.

d_n	mm	Nenndurchmesser	nominal diameter	diamètre nominal	70
d_i	mm	Innendurchmesser	internal diameter	diamètre intérieur	70,0
p_l	$\dfrac{\text{mm } H_2O}{\text{m}}$	verfügbares Druck-gefälle (je 1 m Rohrlänge)	available pressure drop (per metre run of pipe)	chute de pression disponible (par m de conduite)	10,7
w_W	$\dfrac{\text{m}}{\text{s}}$	Strömungs-geschwindigkeit des Wassers	velocity of flow of water	vitesse de circu-lation de l'eau	0,89

Rietschel.

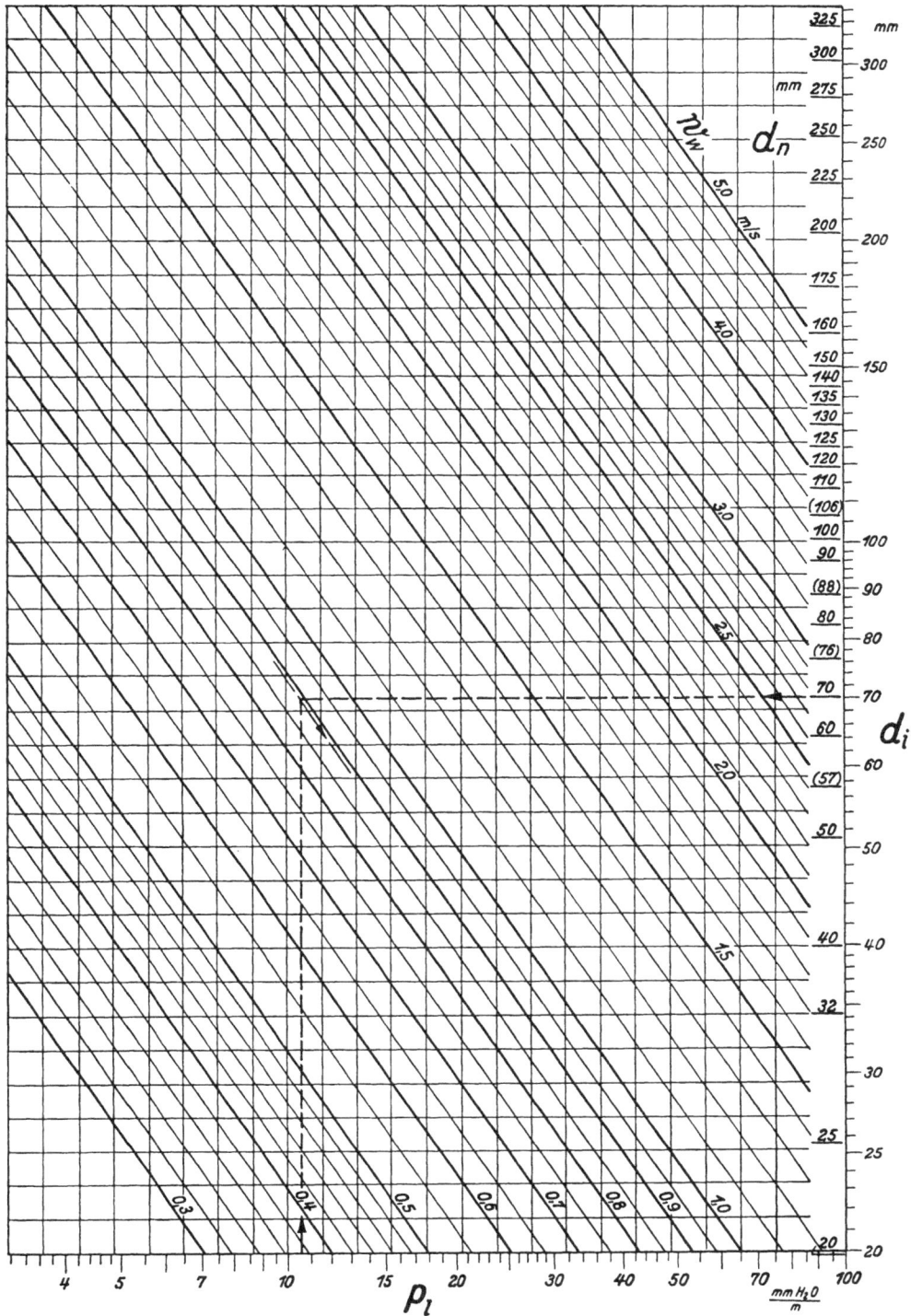

Wärmeleistung von Niederdruck-Dampfheizungen.
Heat Output of Low Pressure Steam Systems.
Pouvoir de chauffe des installations de chauffage par vapeur à basse pression.

d_n	mm	Nenndurchmesser	nominal diameter	diamètre nominal	40
d_t	mm	Innendurchmesser	internal diameter	diamètre intérieur	39,75
Q_h	$\dfrac{\text{kcal}}{\text{h}}$	Wärmeleistung	heat output	quantité de chaleur émise par heure	15000
p_l	$\dfrac{\text{mm } H_2O}{\text{m}}$	Druckgefälle	pressure drop	chute de pression	2,6

Rietschel.

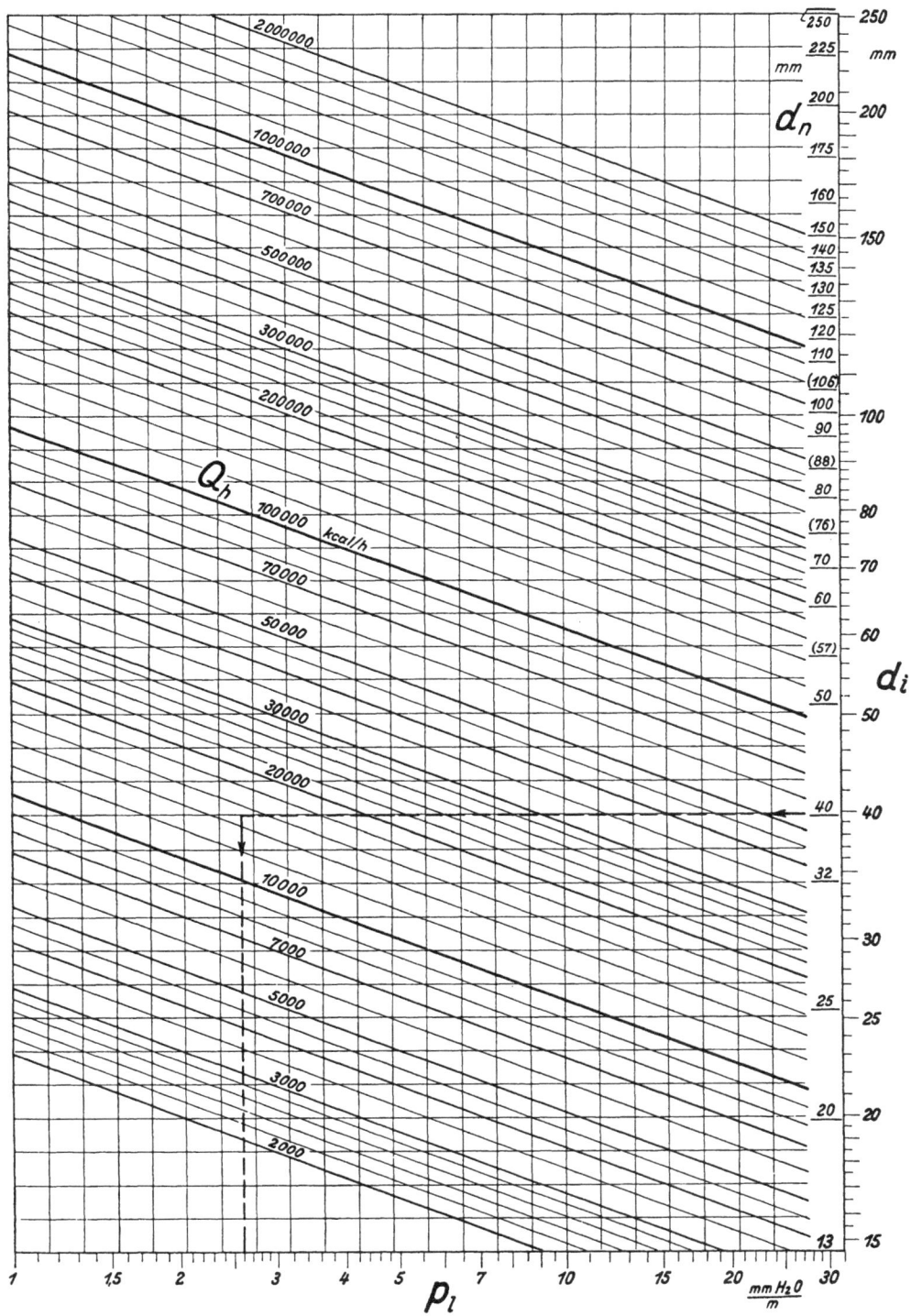

29

Strömungsgeschwindigkeit in Niederdruck-Dampfheizungen.
Velocity of Flow in Low Pressure Steam Systems.
Vitesse de Circulation dans les installations de chauffage par vapeur à basse pression.

d_n	mm	Nenndurchmesser	nominal diameter	diamètre nominal	40
d_i	mm	Innendurchmesser	internal diameter	diamètre intérieur	39,75
p_i	$\dfrac{\text{mm } H_2O}{\text{m}}$	Druckgefälle	pressure drop	chute de pression	2,6
w_D	$\dfrac{\text{m}}{\text{s}}$	Strömungs-geschwindigkeit des Dampfes	velocity of flow of steam	vitesse d'écoule-ment de la va-peur	9,7

Rietschel.

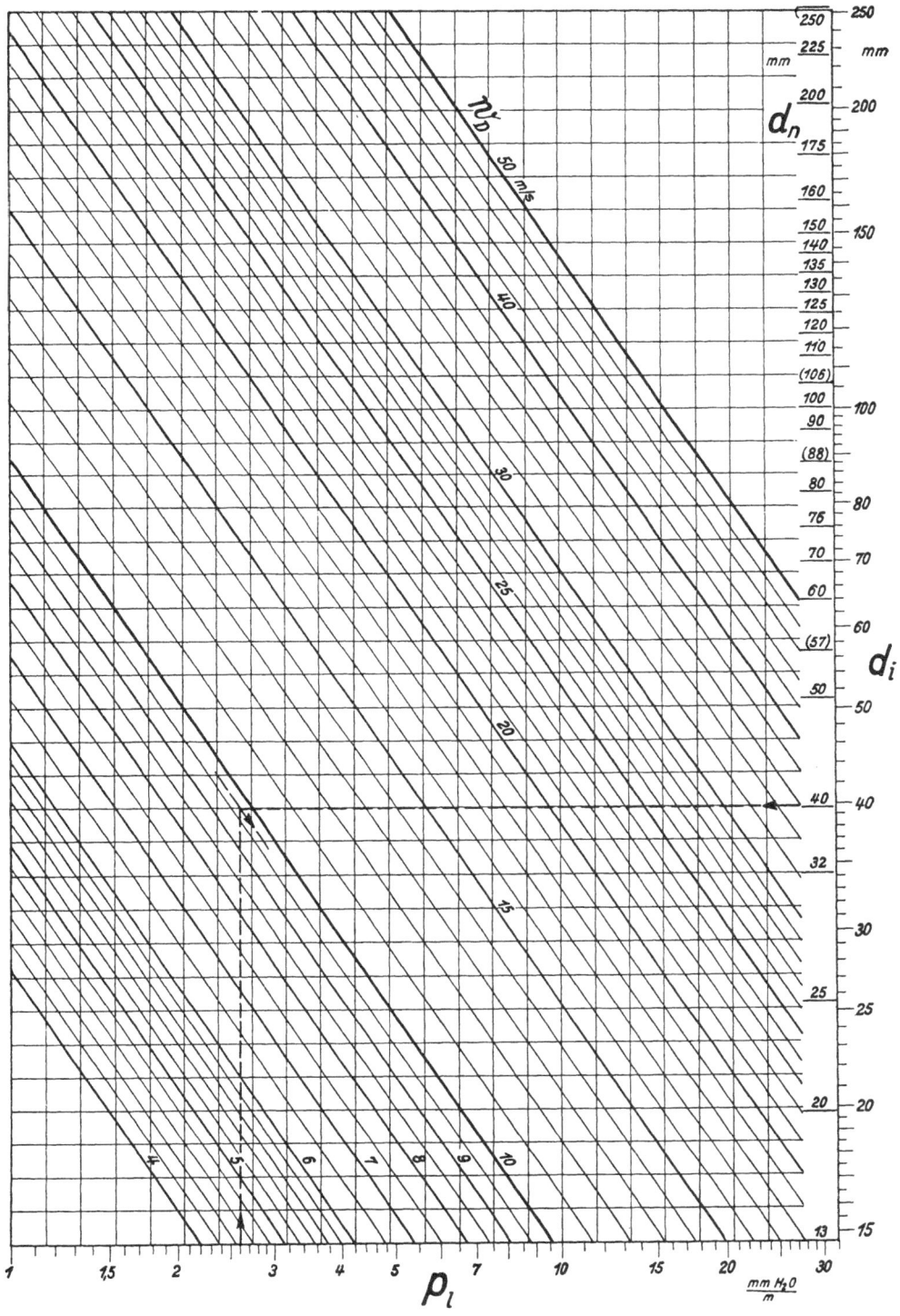

30

Kondenswasserleitungen.
Condensate Pipes.
Conduites à eau condensée.

						①	②
d_n	mm	Nenndurchmesser	nominal diameter	diamètre nominal		60	60
d_i	mm	Innendurchmesser	internal diameter	diamètre intérieur		64,0	64,0
		Leitungsart	pipe arrangement	disposition de la canalisation		HL–V	TL
l_{max}	m	Länge der Rohrleitung (zum Heizkörper, der vom Kessel am weitesten entfernt ist)	length of pipe line (to radiator furthest from boiler)	longueur de la canalisation (pour le radiateur le plus éloigné de la chaudière)			70
Q_D	$\dfrac{1000 \text{ kcal}}{h}$	Wärmemenge im niedergeschlagenen Dampf	heat content of condensed steam	chaleur contenue dans la vapeur condensée		635	850

Leitungsarten. — Pipe Arrangements. — Disposition des canalisations.

HL	hochliegende Leitungen	high level	sous plafond
TL	tiefliegende Leitungen	low level	en dessous
H	waagerechte Leitungen	horizontal	horizontales
V	lotrechte Leitungen	vertical	verticales

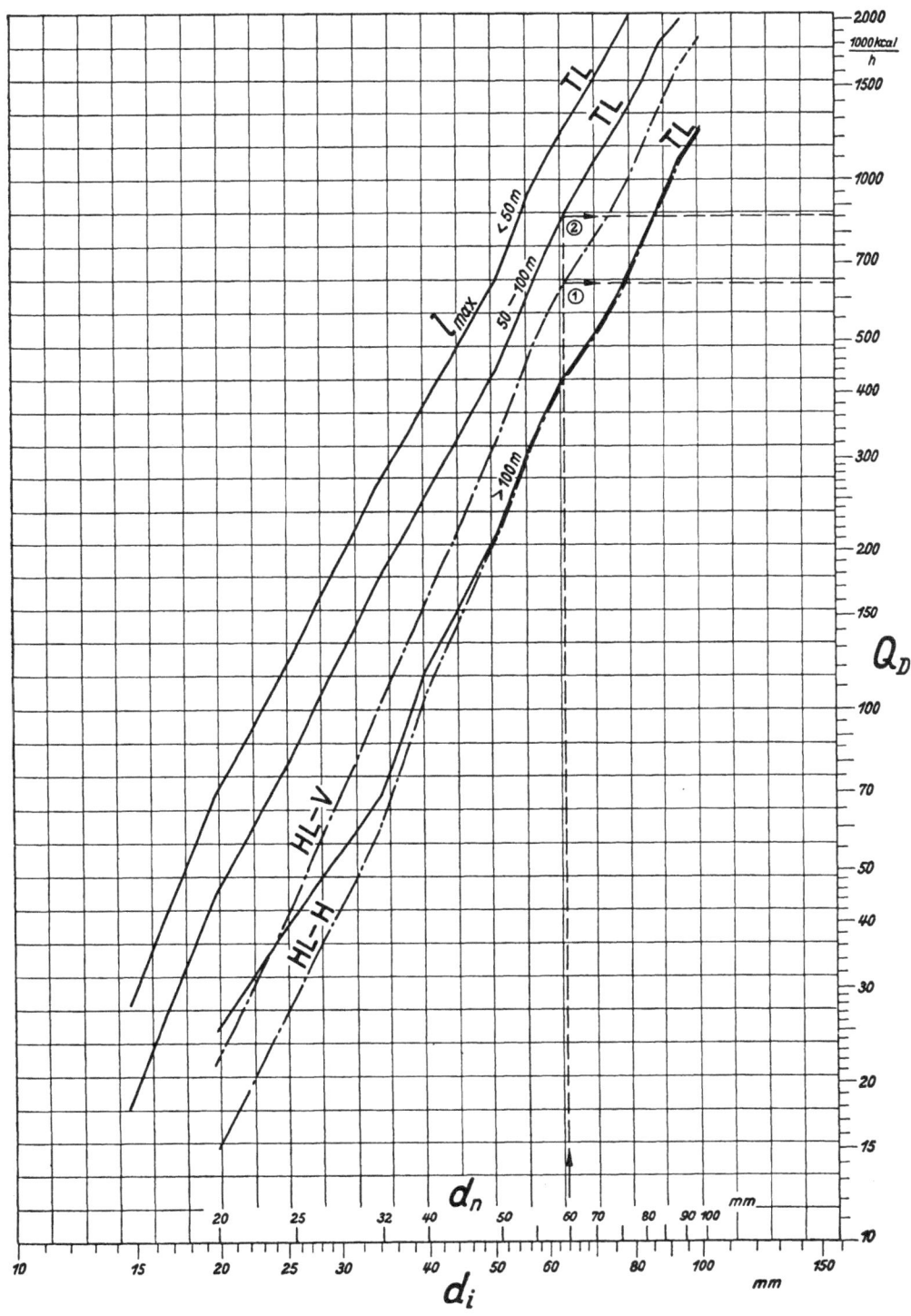

Wärmedurchgang für Heizflächen.
Heat Transmission of Radiators.
Transmission de chaleur par les radiateurs.

①

E	mm	Nabenabstand	distance between bosses	entraxe des orifices	555
		Heizmittel	heating medium	agent de chauffage	W
		Heizkörperart	type of heater	type de radiateur	Lr
C	mm	Tiefe des Heiz-körpers	depth of radiator	saillie du radiateur	190
k	$\dfrac{kcal}{m^2\,h\,{}^0C}$	Wärmedurch-gangszahl	coefficient of heat transmission	coefficient de trans-mission	6,65
R	$\dfrac{m^2\,h\,{}^0C}{kcal}$	Wärmewiderstand	thermal resistance	résistance ther-mique	0,15

②

E	mm	Nabenabstand	distance between bosses	entraxe des orifices	700
		Heizmittel	heating medium	agent de chauffage	W
		Heizkörperart	type of heater	genre de radiateur	Nr
S	mm	Anzahl der Säulen des Heizkörpers	number of radiator columns	nombre de colonnes du radiateur	3
k	$\dfrac{kcal}{m^2\,h\,{}^0C}$	Wärmedurch-gangszahl	coefficient of heat transmission	coefficient de trans-mission	6,2

③

d_i	mm	Innendurchmesser	internal diameter	diametre intérieur	64,0
		Heizmittel	heating medium	agent de chauffage	D
		Heizkörperart	type of heater	genre de radiateur	RH—M
k	$\dfrac{kcal}{m^2\,h\,{}^0C}$	Wärmedurch-gangszahl	coefficient of heat transmission	coefficient de trans-mission	9,6

Heizmittelarten. — Heating Medium. — Agents de chauffage.

| W | Warmwasser | hot water | eau chaude |
| D | Niederdruckdampf | low pressure steam | vapeur à basse pression |

Heizkörperarten. — Type of Heater. — Genres de radiateur.

NR	Normalradiator	standard radiator	radiateur normal
LR	Leichtradiator	light radiator	radiateur léger
RH	Rohrheizkörper	tubular heater (smooth horizon-tal pipes)	faisceau tubulaire horizon-tal, à tubes lisses
$RR-E$	Rippenrohre, ein-zeln	ribbed pipes, single	tubes à ailettes
$RR-M$	Rippenrohre, mehr-fach übereinander	ribbed pipes, mul-tiple (one above another)	batteries de tubes à ailet-tes superposés

DIN 4701. — Rietschel.

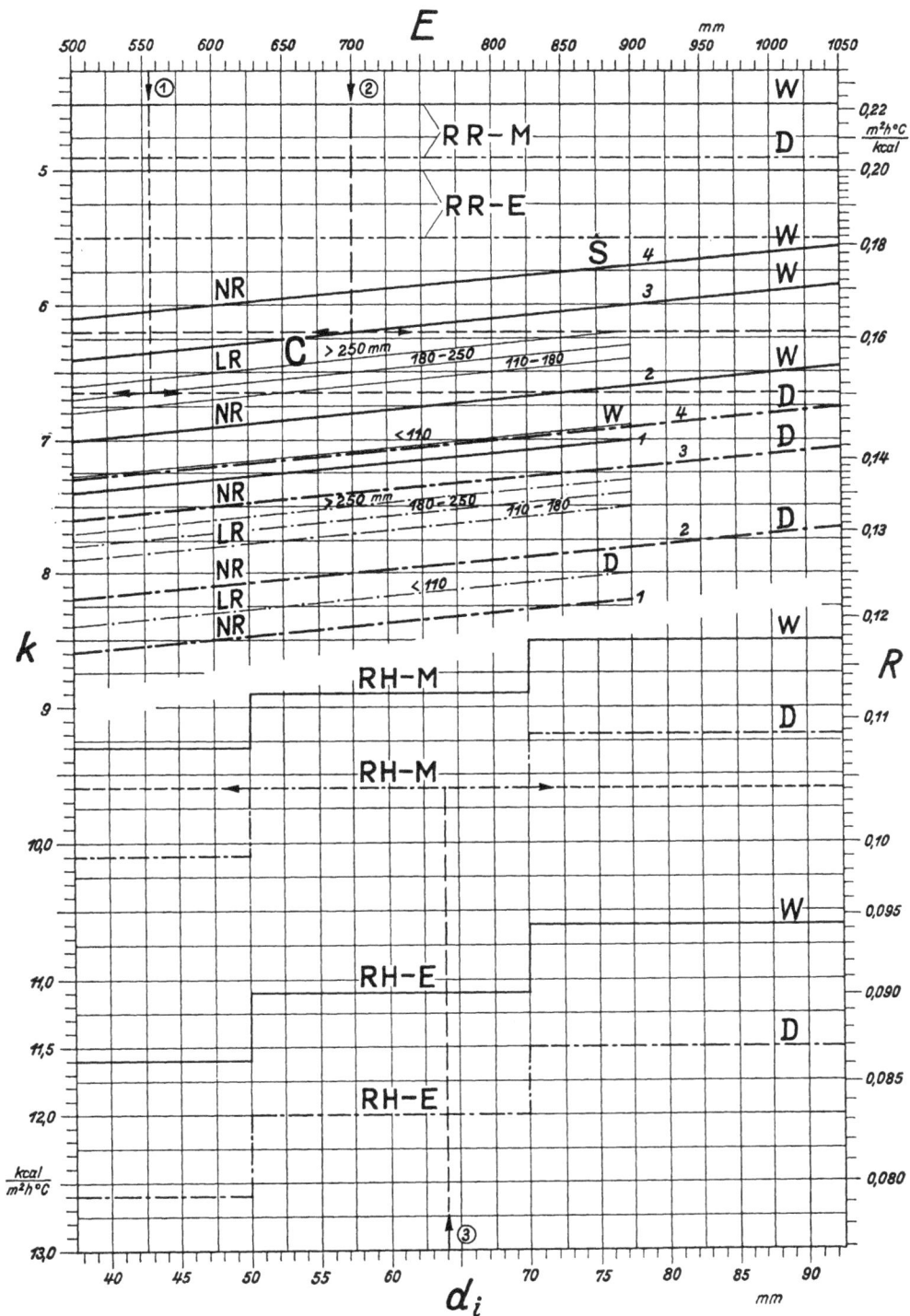

32

Wärmeabgabe nackter Rohre.
Heat Emission from Bare Pipes.
Pouvoir émissif des tuyauteries nues.

d_n	mm	Nenndurchmesser	nominal diameter	diamètre nominal	100
d_i	mm	Innendurchmesser	inside diameter	diamètre intérieur	100,5
Δt	°C	Temperaturunter-schied	temperature differ-ence	différence de tem-pérature	40
q_r	$\dfrac{\text{kcal}}{\text{m h}}$	Wärmeabgabe (je 1 m Rohrlänge)	heat emission per metre run of pipe line	chaleur émise par m de conduite	115

Recknagel.

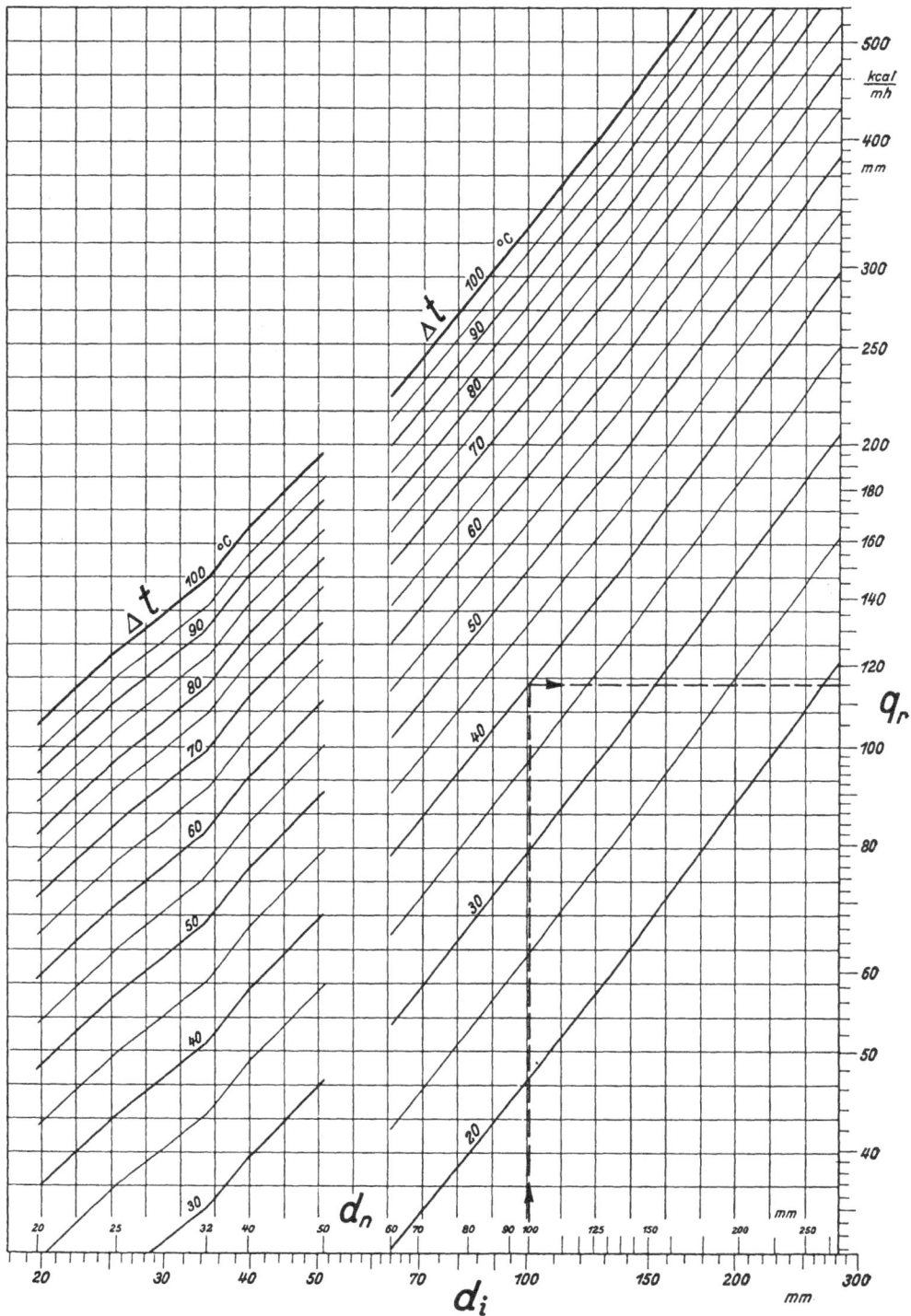

33

Wärmedurchgang durch Kupfer- und Eisenrohre.
Heat Transmission by Copper and Iron Pipes.
Transmission de chaleur à travers les tubes en cuivre et en fer.

					①	②
$w_D w_W$	$\frac{m}{s}$	Strömungs-geschwindigkeit des Dampfes bzw. Wasser	velocity of flow of steam or water	vitesse de circu-latiou de la vapeur ou de l'eau	0,75	
		Heizmittel	heating medium	agent de chauffage	D	W
		Rohrwerkstoff	pipe material	matière des tuyaux	Cu	Fe
k	$\frac{kcal}{m^2\,h\,°C}$	Wärmedurch-gangszahl	coefficient of heat transmission	coefficient de transmission	1530	1390

Heizmittelarten. — Heating Medium. — Agents de chauffage.

W	Warmwasser	hot water	eau chaude
D	Niederdruckdampf	low pressure steam	vapeur à basse pres-sion

Rohrwerkstoffarten. — Pipe materials. — Matière des tubes.

Cu	Kupferrohre	copper pipes	tubes en cuivre
Fe	Eisenrohre	iron pipes	tubes en fer

Recknagel.

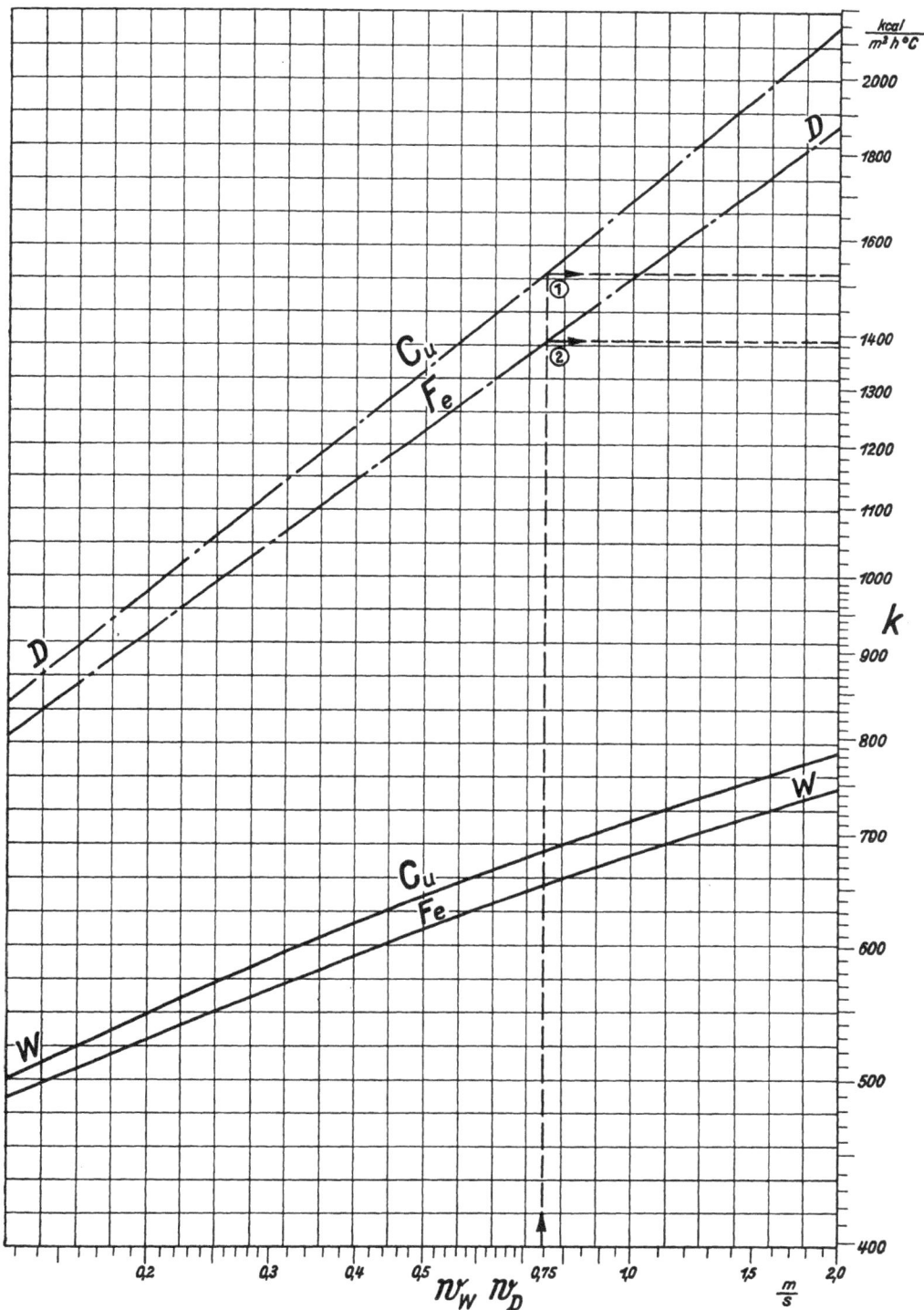

34

Mittlerer Temperaturunterschied (in Wärmeaustauschern).
Mean Temperature Difference (in Heat Exchanger Apparatus).
Ecart moyen de température (dans les échangeurs de chaleur).

					①	②
Δt_1	°C	Temperaturunterschied am Anfang der Heizfläche	temperature difference at beginning of heating surface	écart de température au début de la surface de chauffe	94	1150
Δt_2	°C	Temperaturunterschied am Ende der Heizfläche	temperature difference at end of heating surface	écart de température à la fin de la surface de chauffe	30	200
Δt_m	°C	mittlerer Temperaturunterschied	mean temperature difference	écart moyen de température	56	540

$$\Delta t_m = \frac{\Delta t_1 - \Delta t_2}{\ln \Delta t_1 - \ln \Delta t_2}$$

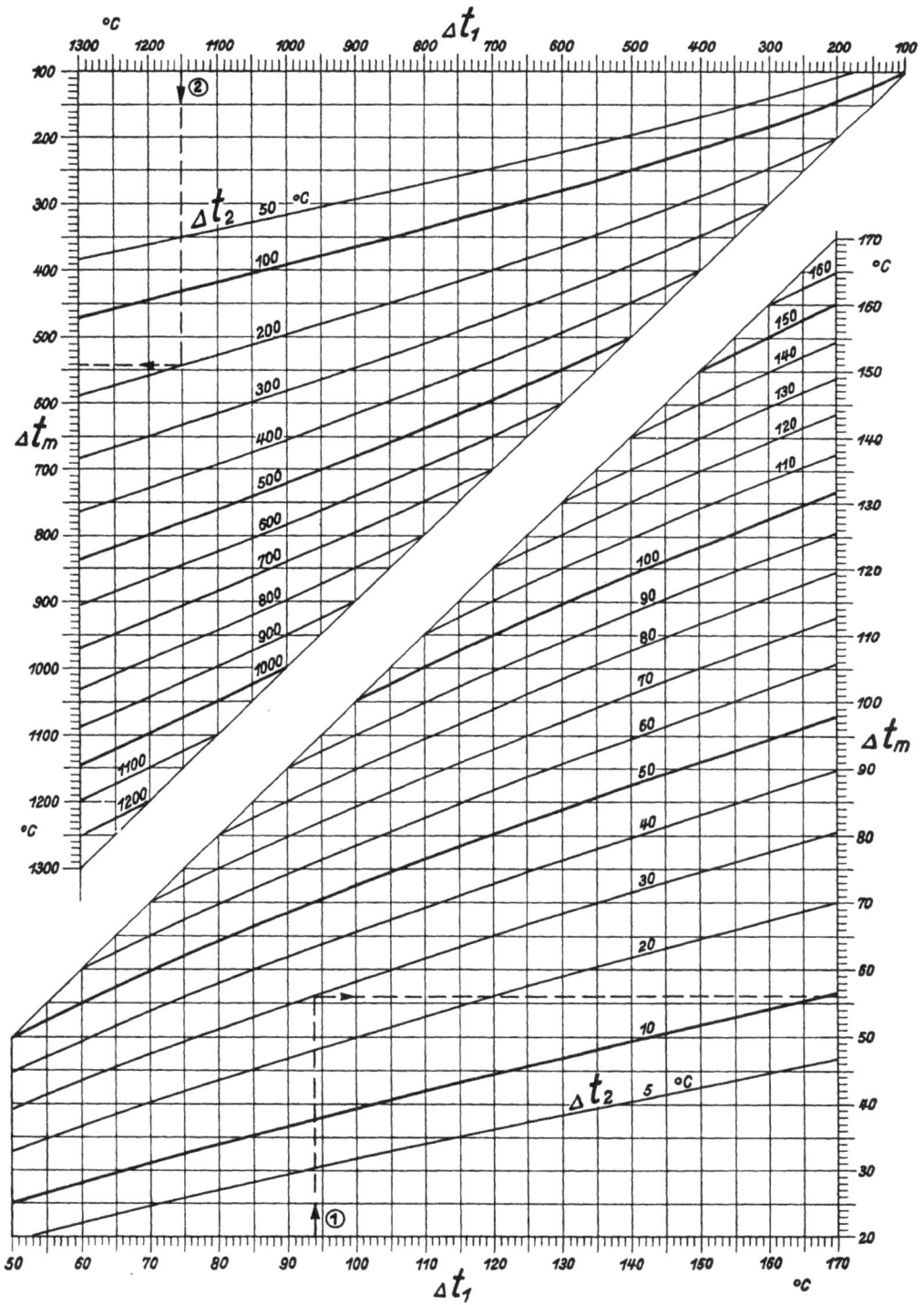

35

Regelung von Warmwasserheizungen.
Control of Hot Water Systems.
Réglage des installations de chauffage à eau chaude.

		Heizungsart	heating system	installation de chauffage	① S	② P
t_a	°C	Außentemperatur	outdoor temperature	température extérieure	—7,0	—7,0
t_v	°C	Wassertemperatur im Vorlauf	temperature of water in flow	température de l'eau dans la canalisation d'amenée	72,8	71,4
t_m	°C	mittlere Wassertemperatur	mean temperature of water	température moyenne de l'eau	64,7	64,7
t_r	°C	Wassertemperatur im Rücklauf	temperature of water in return	température de l'eau dans la canalisation de retour	56,6	58,0
$t_v - t_r$	°C	Temperaturunterschied zwischen Vor- und Rücklauf	temperature difference between flow and return	différence de température entre les canalisations d'amenée et de retour	16,2	13,5

Arten der Warmwasserheizungen. — Hot Water Heating Systems. — Genres d'installations de chauffage.

S	Schwerkraft	gravity systems	installations à thermo-siphon
P	Pumpen	forced circulation systems	installations avec circulation par pompe

Barenbrug, Berechnung der Vor- und Rücklauf-Temperaturen in Abhängigkeit von den Außentemperaturen. Gesundh.-Ing. Bd. 57 (1934), S. 425.

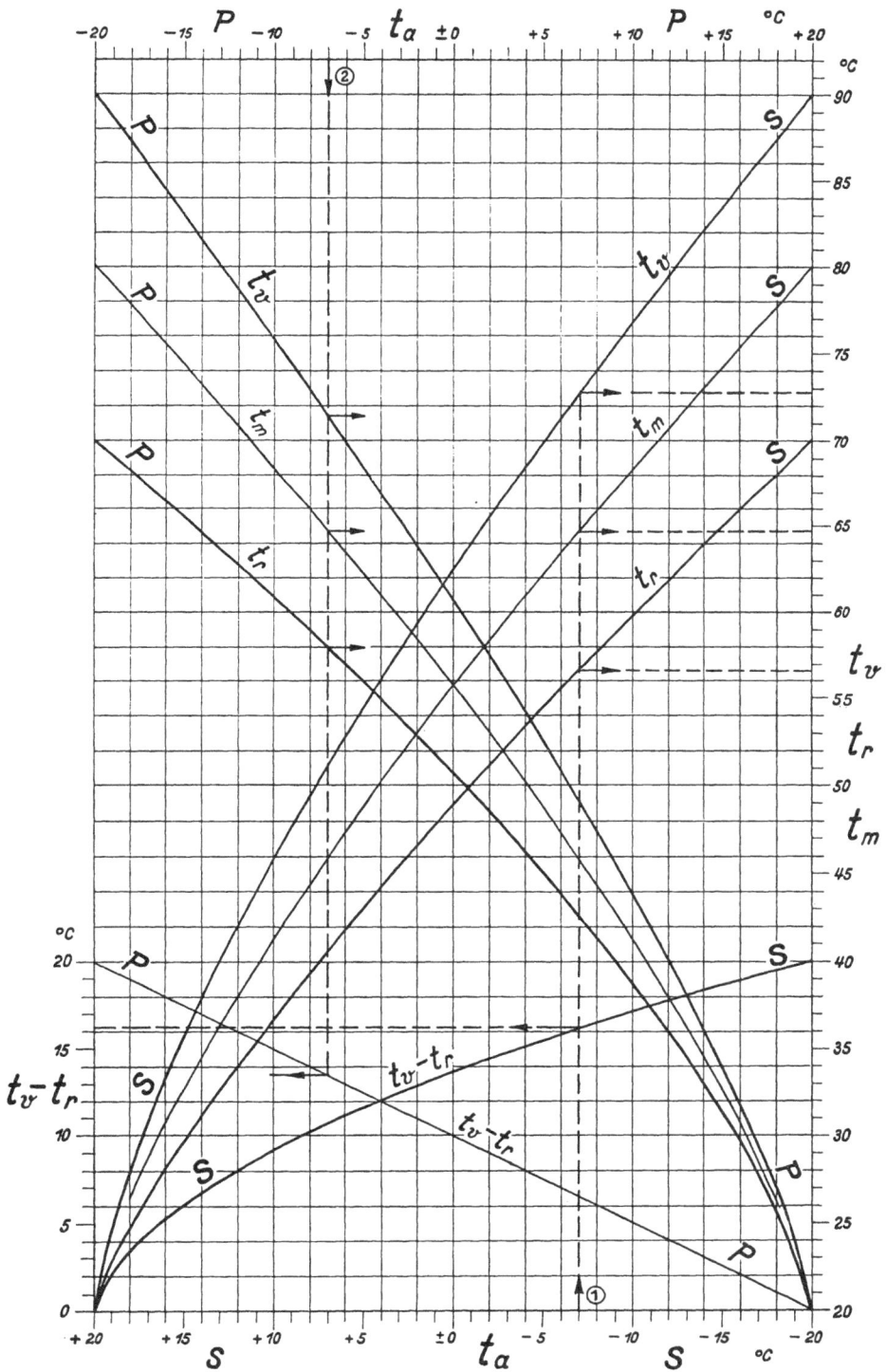

36

Brennstoffverbrauch.
Fuel Consumption.
Consommation de combustible.

q_G	$\dfrac{1000\ \text{kcal}}{°C\ (24\ h)}$	Wärmeverbrauch (je 1 Gradtag)	heat consumption per degree Centigrade per 24 hr. (i. e. per degree-day)	consommation de chaleur (par degré et par jour)	60
η_h	$°/_0$	Wirkungsgrad der gesamten Heizanlage	overall efficiency of heating installation	rendement global de l'installation de chauffage	69
ε		Heizkennziffer			1,45
G	$\dfrac{°C\ (24\ h)}{a}$	Gradtagzahl (im Jahr)	number of degree days (per year)	nombre de jours-degrés (par an)	2200
H_u	$\dfrac{\text{kcal}}{\text{kg}}$	unterer Heizwert	net calorific value	valeur calorifique inférieur	7000
B_a	$\dfrac{t}{a}$	jährlicher Brennstoffverbrauch	annual consumption of fuel	consommation annuelle de combustible	27,5

$$B_a = \frac{q_G \cdot G \cdot \varepsilon}{H_u}$$

$$\varepsilon = \frac{100}{\eta_h}$$

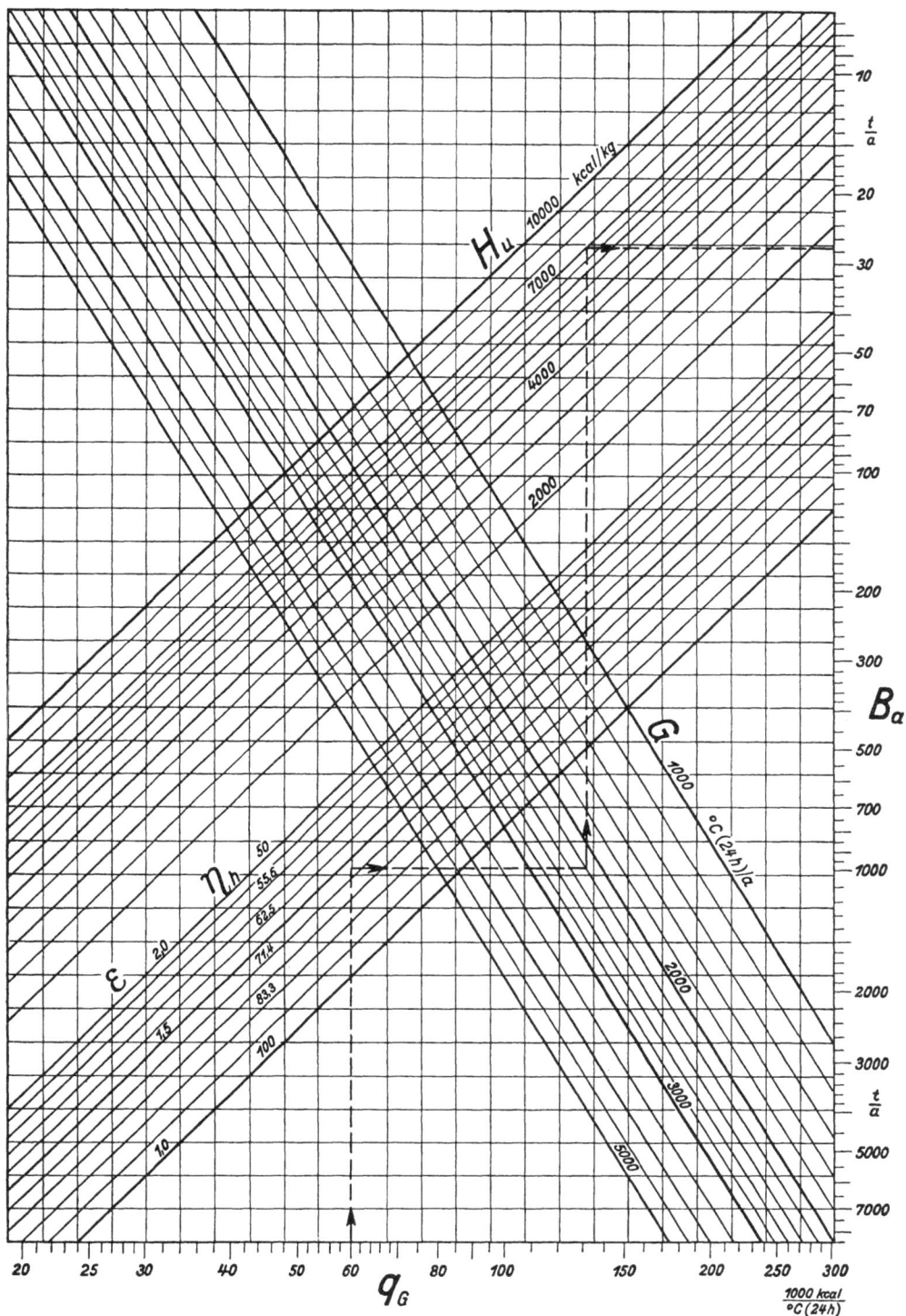

37

Brennstoffeigenschaften.
Properties of Fuels.
Propriétés des combustibles.

		Brennstoffart	kind of fuel	genre de combustible	wS
		Gewichtsanteil:	percentage by	teneur en poids:	
$g[S]$	$\%$	des Schwefels	weight of: sulphur	en soufre	1
$g[ON]$	$\%$	des Sauerstoffs und Stickstoffs	oxygen and nitrogen	oxygène et azote	7
$g[H]$	$\%$	des Wasserstoffs	hydrogen	hydrogène	5
$g[C]$	$\%$	des Kohlenstoffs	carbon	carbone	79
w	$\%$	Wassergehalt des Brennstoffs	moisture content of fuel	teneur en eau du combustible	3
a	$\%$	Aschengehalt des Brennstoffs	ash content of fuel	teneur en cendres du combustible	5
γ_B	$\dfrac{kg}{m^3}$	spez. Gewicht des Brennstoffs	density of fuel	poids spécifique du combustible	1350
H_u	$\dfrac{kcal}{kg}$	unterer Heizwert	net calorific value	pouvoir calorifique inférieur	7620
H_o	$\dfrac{kcal}{kg}$	oberer Heizwert	gross calorific value	pouvoir calorifique supérieur	7860

Brennstoffarten. — Kinds of Fuel. — Genres de combustible.

H	Holz	wood	bois
T	Torf	peat	tourbe
lB	Lausitzer Braunkohle	Lausitz brown coal	lignite de Lusace
bB	Böhmische Braunkohle	Bohemian brown coal	lignite de Bohême
sS	schlesische Steinkohle	Silesian hard coal	houille de Silésie
wS	westfälische Steinkohle	Westphalian hard coal	houille de Westphalie
A	Anthrazit	anthracite	anthracite
BB	Braunkohlenbriketts	lignite briquette	briquette de lignite
tK	trockener Koks	dry coke	coke sec
fK	feuchter Koks	wet coke	coke humide
GO	Gasöl	gas-oil	gas oil
BO	Braunkohlenteeröl	brown-coal tar oil	huile de goudron de lignite
SO	Steinkohlenteeröl	hard-coal tar oil	huile de goudron de houille

Rietschel.

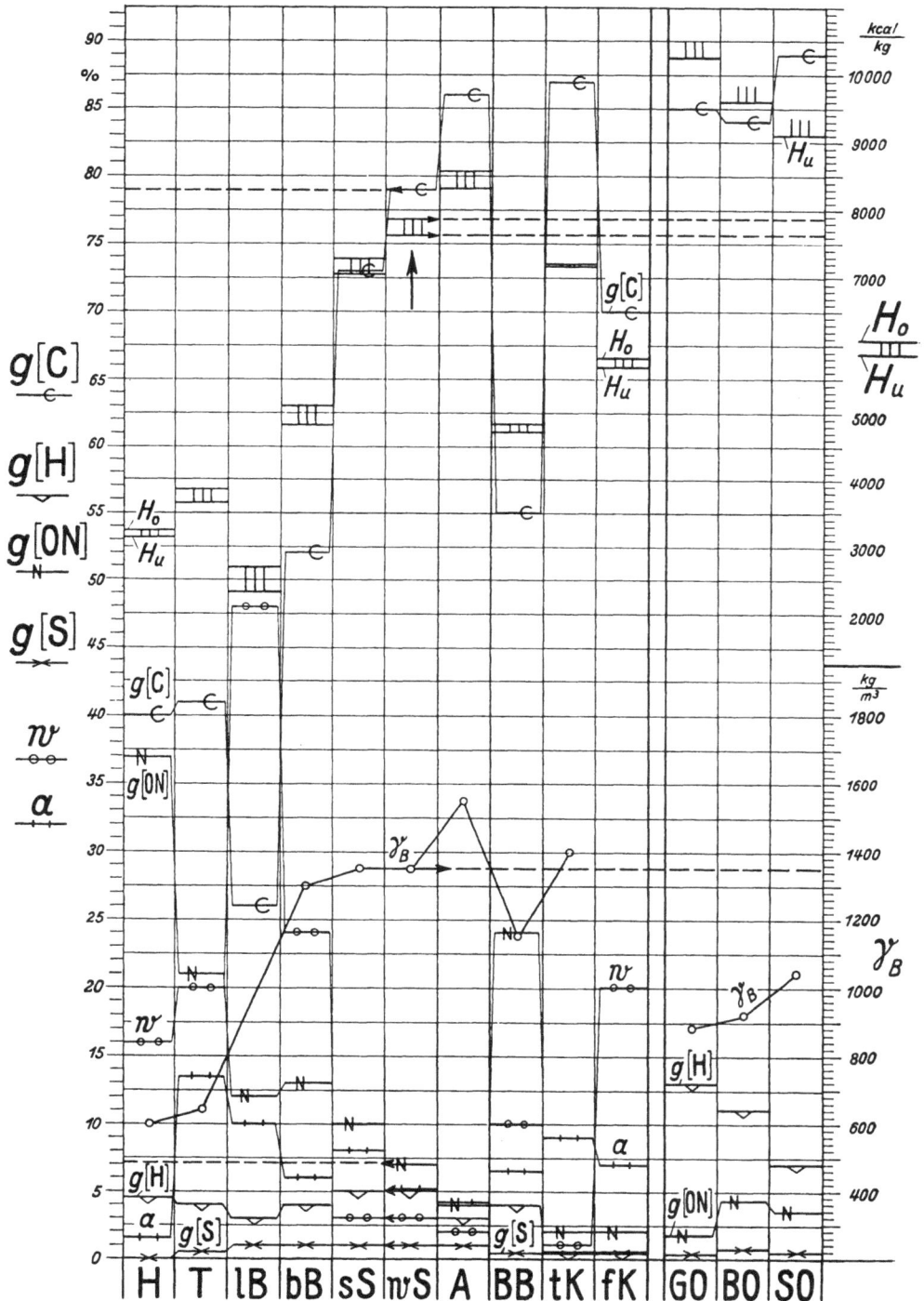

Koks-Korngröße.
Size of Coke.
Grosseur du coke.

h_g	cm	Glutschichthöhe	thickness of firebed	hauteur de la couche en ignition	42
F_k	m²	Kesselheizfläche	boiler heating surface	surface de chauffe de la chaudière	18
l_K	mm	Koks-Korngröße	size of coke	grosseur du coke	67
		Bezeichnung für Ruhrkoks	designation for Ruhr coke	désignation cemmerciale (coke de la Ruhr)	I

Schmidt, Rainer-Schmidt, Wahl und Abnahme der richtigen Kokssorte für Zentralheizungen. Gesundh.-Ing. Bd. 56 (1933), S. 373.

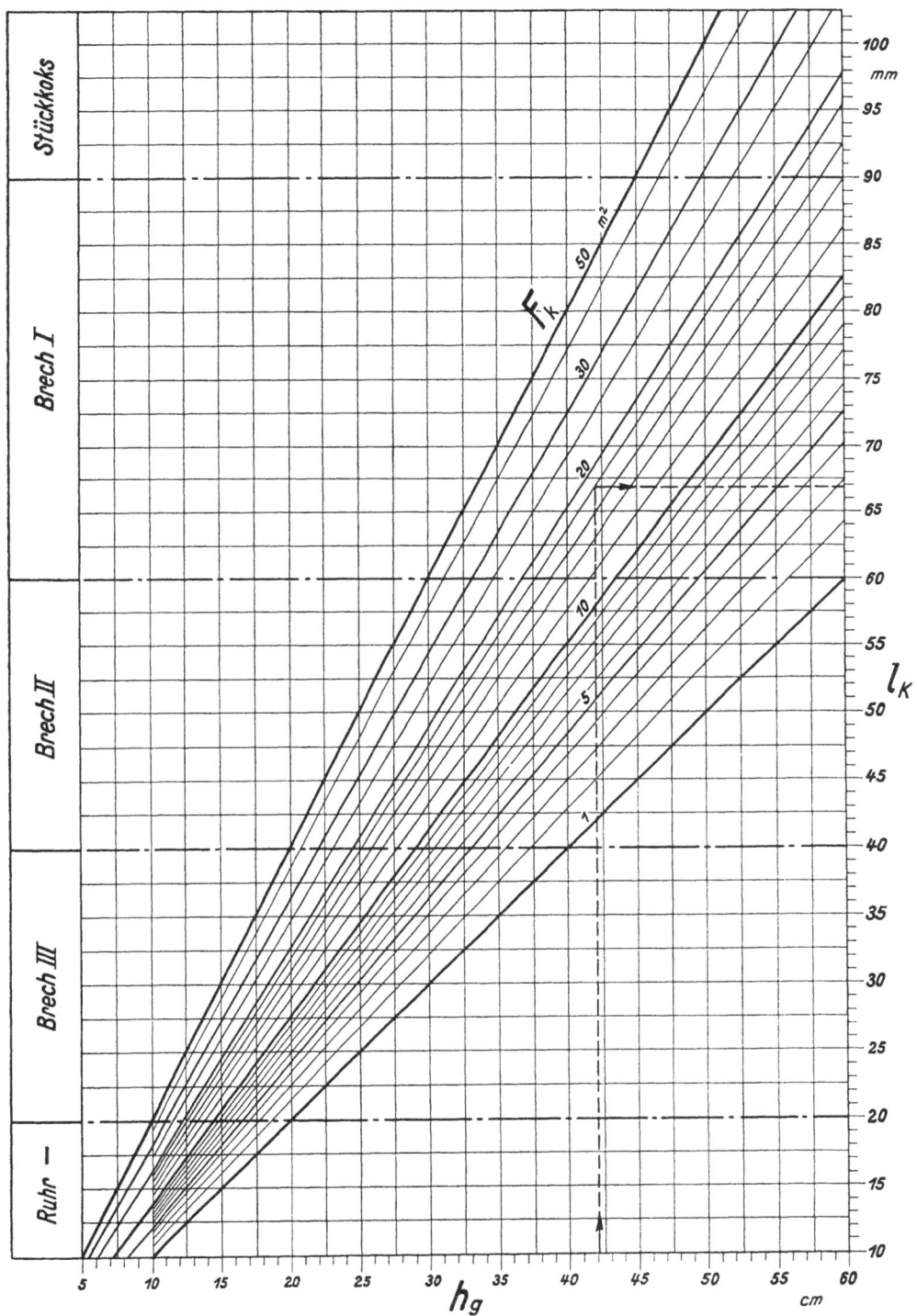

39

Brennstoff- und Wärmekosten.
Fuel Costs and Heat Costs.
Coût du combustible et coût de la chaleur.

k_B	$\dfrac{M}{t}$	Brennstoffkosten	price of fuel (per ton)	prix du combustible (par tonne)	45
H_u	$\dfrac{kcal}{kg}$	unterer Heizwert	net calorific value	pouvoir calorifique inférieur	7000
k_Q	$\dfrac{M}{10^6\,kcal}$	Wärmekosten (im Brennstoff)	cost of heat in the fuel	coût de la chaleur dans le combustible	6,43
η_h	$^0/_0$	Gesamtwirkungsgrad der Heizanlage	overall efficiency of heating installation	rendement global de l'installation de chauffage	69
k_{Qh}	$\dfrac{M}{10^6\,kcal}$	Wärmekosten (im beheizten Raum)	cost of heat in the heated space	coût de la chaleur dans le local chauffé	9,35

$$k_Q = \frac{1000 \cdot k_B}{H_u}$$

$$k_{Q_h} = \frac{k_Q}{\eta_h}$$

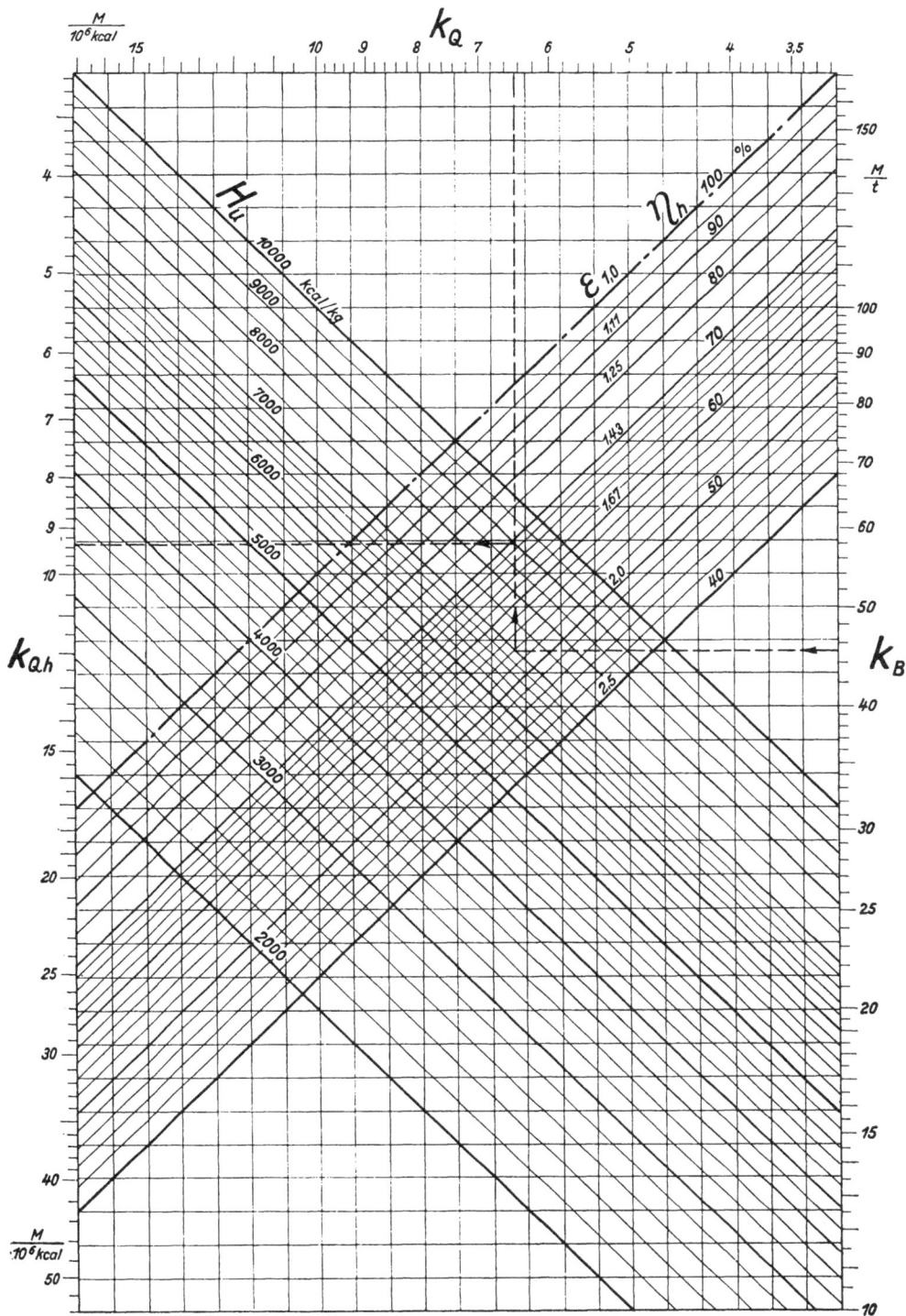

40

Wärmeeigenschaften von Warmwasser und Niederdruckdampf.
Thermal Properties of Hot Water and Low Pressure Steam.
Propriétés thermiques de l'eau chaude et de la vapeur d'eau saturée.

I. Warmwasser. — Hot Water. — Eau chaude.

①

t_W	°C	Wassertemperatur	temperature of water	température de l'eau	84
p_D	ata	Sättigungsdruck	saturated steam pressure	pression de la vapeur	0,56
i_W	$\dfrac{kcal}{kg}$	Wärmeinhalt des Wassers	heat content of water	chaleur contenue dans l'eau	84
γ_W	$\dfrac{kg}{m^3}$	spezifisches Gewicht des Wassers	density of water	poids spécifique de l'eau	969,4

II. Niederdruckdampf (Sattdampf). — Low Pressure Steam (Saturated). — Vapeur d'eau saturée.

p_D	ata	Dampfdruck	steam pressure	pression de la vapeur	1,4
t_W	°C	Sättigungstemperatur	saturation temperature	température de saturation	109
i_r	$\dfrac{kcal}{kg}$	Verdampfungswärme	latent heat of evaporation	chaleur latente d'évaporation	534
γ_D	$\dfrac{kg}{m^3}$	spezifisches Gewicht des Dampfes	density of steam	poids spécifique de la vapeur	0,79
i_D	$\dfrac{kcal}{kg}$	Wärmeinhalt des Dampfes	heat content of steam	chaleur de la vapeur	643

Knoblauch, Raisch, Hausen, Koch, Tabellen und Diagramme für Wasserdampf.
München 1932.

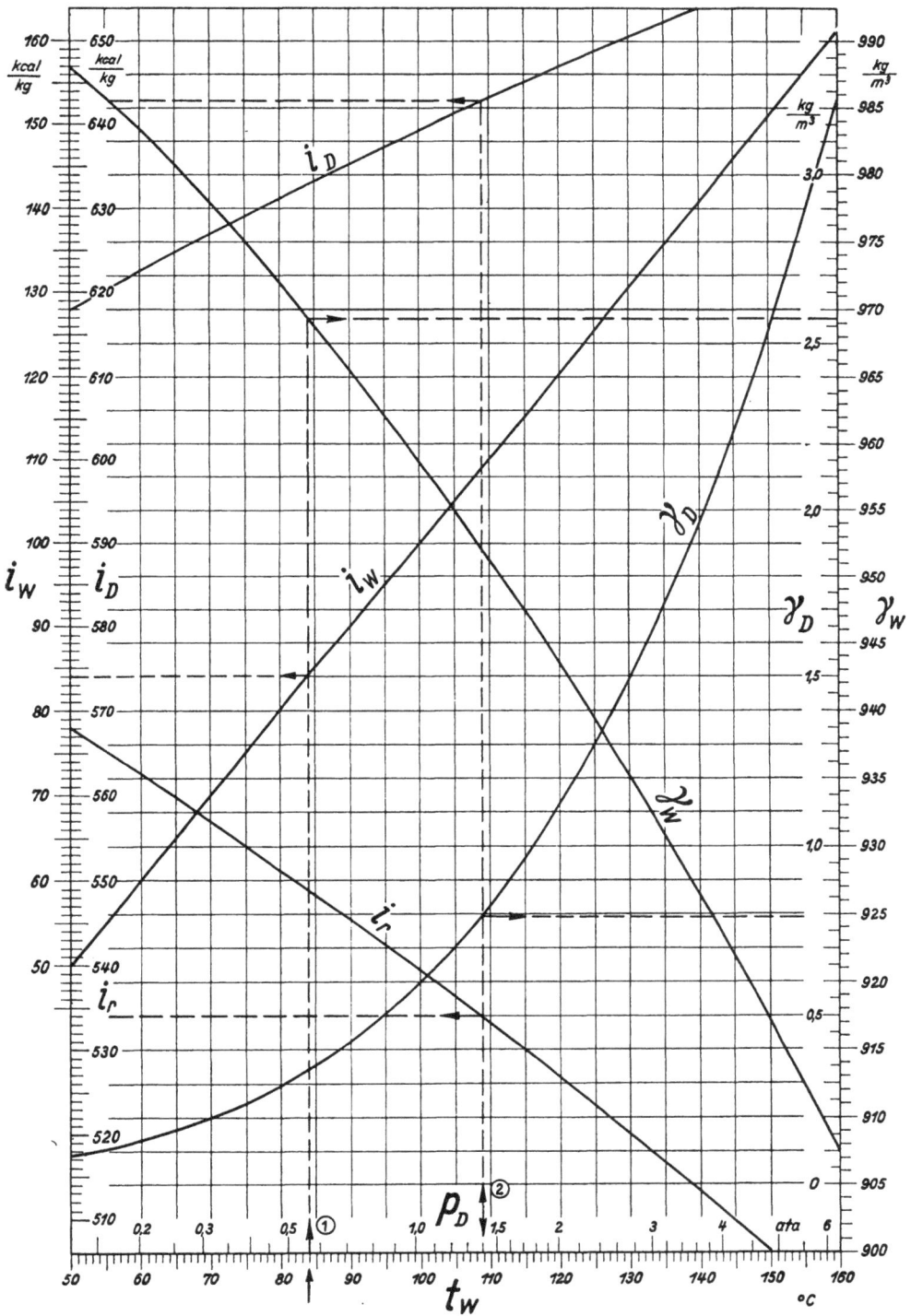